# Deep Learning and Reinforcement Learning

*Edited by Jucheng Yang, Yarui Chen, Tingting Zhao, Yuan Wang and Xuran Pan*

Published in London, United Kingdom

Deep Learning and Reinforcement Learning
http://dx.doi.org/10.5772/intechopen.103984
Edited by Jucheng Yang, Yarui Chen, Tingting Zhao, Yuan Wang and Xuran Pan

Contributors
Ning Weng, Prashant Baral, Ning Yang, Bhanu K.N. Prakash, Arvind Channarayapatna Srinivasa, Ling Yun Yeow, Wen Xiang Chen, Audrey Jing Ping Yeo, Wee Shiong Lim, Cher Heng Tan, Hany Mohamed Nabil Helmy, Sherif El Diasty, Hazem Shatila, Yuan Wang, Zekun Li, Zhenyu Deng, Huiling Song, Jucheng Yang, Fateme Fathinezhad, Jocelyn Chanussot, Peyman Adibi, Bijan Shoushtarian, Ramzi Mahmoudi, Narjes Ben Ameur



First published in London, United Kingdom, 2023 by IntechOpen
IntechOpen is the global imprint of INTECHOPEN LIMITED, registered in England and Wales, registration number: 11086078, 5 Princes Gate Court, London, SW7 2QJ, United Kingdom
Printed in Croatia

British Library Cataloguing-in-Publication Data
A catalogue record for this book is available from the British Library

Additional hard and PDF copies can be obtained from orders@intechopen.com

Deep Learning and Reinforcement Learning
Edited by Jucheng Yang, Yarui Chen, Tingting Zhao, Yuan Wang and Xuran Pan
p. cm.

This title is part of the Artificial Intelligence Book Series, Volume 18
Topic: Machine Learning and Data Mining
Series Editor: Andries Engelbrecht
Topic Editor: Marco Antonio Aceves-Fernández

Print ISBN 978-1-80356-950-5
Online ISBN 978-1-80356-951-2
eBook (PDF) ISBN 978-1-80356-952-9
ISSN 2633-1403

For EU product safety concerns:
IN TECH d.o.o., Prolaz Marije Krucifikse Kozulić 3, 51000 Rijeka, Croatia,
info@intechopen.com or visit our website at intechopen.com.

# IntechOpen Book Series
# Artificial Intelligence
## Volume 18

## Aims and Scope of the Series

Artificial Intelligence (AI) is a rapidly developing multidisciplinary research area that aims to solve increasingly complex problems. In today's highly integrated world, AI promises to become a robust and powerful means for obtaining solutions to previously unsolvable problems. This Series is intended for researchers and students alike interested in this fascinating field and its many applications.

# Meet the Series Editor

Andries Engelbrecht received the Masters and Ph.D. degrees in Computer Science from the University of Stellenbosch, South Africa, in 1994 and 1999 respectively. He is currently appointed as the Voigt Chair in Data Science in the Department of Industrial Engineering, with a joint appointment as Professor in the Computer Science Division, Stellenbosch University. Prior to his appointment at Stellenbosch University, he has been at the University of Pretoria, Department of Computer Science (1998-2018), where he was appointed as South Africa Research Chair in Artifical Intelligence (2007-2018), the head of the Department of Computer Science (2008-2017), and Director of the Institute for Big Data and Data Science (2017-2018). In addition to a number of research articles, he has written two books, Computational Intelligence: An Introduction and Fundamentals of Computational Swarm Intelligence.

# Meet the Volume Editors

Jucheng Yang is a Distinguished Professor at the College of Artificial Intelligence, Tianjin University of Science and Technology, China. He was a visiting professor at the University of New South Wales, Australia, and the University of Surrey, UK. He has published more than 140 papers in international journals and conferences. He has served as an associate editor of *Frontiers in Plant Science and Journal of Information Security and Applications* and a reviewer of international journals such as *IEEE Transactions on Information Forensics and Security, IEEE Transactions on Industrial Informatics*, and *IEEE Communications Magazine*. He has served as the chairman of eight conferences and a member of the program committees of more than thirty conferences. His research interests include biometrics, image processing, neural networks, and information security.

Yarui Chen is a professor at the College of Artificial Intelligence, Tianjin University of Science and Technology, China. She received her BS from Hebei University of Technology, China, and an MS and Ph.D. from Tianjin University, China. She has published more than twenty papers in journals and conferences. She has served as a reviewer of international journals such as *Neurocomputing*, *KSII Transactions on Internet and Information Systems*, *Journal of Computer Research and Development*, and others. Dr. Chen has three Chinese biometric patents to her credit. Her research interests include machine learning, neural networks, probabilistic inference, and approximate inference.

Tingting Zhao is an Associate Professor in the College of Artificial Intelligence, Tianjin University of Science and Technology, China. She obtained a Ph.D. in Computer Science from Tokyo Institute of Technology, Japan, in 2014. She has authored one book and published more than fifty papers in leading international journals and conferences on machine learning, including neural computation, neural networks and neural information processing systems. She holds more than ten patents. She has served as a program member of top conferences. She has wide research interests, including machine learning and data mining, with a focus on reinforcement learning, robot control, and intelligence systems.

Yuan Wang is an Associate Professor at the College of Artificial Intelligence, Tianjin University of Science and Technology, Tianjin, China. She obtained her Ph.D. in Computer Application Technology at Nankai University, Tianjin, China. She attended the University of California, Santa Cruz, USA, as a visiting scholar in 2014–2015. Her research interests include machine learning, pattern recognition, natural language processing, data mining, and smart cities. She has hosted and participated in several research projects of the National Natural Science Foundation. Dr. Wang has published more than seventy papers in refereed proceedings and journals. She won first place in the shared task of "Sentiment Classification with Deep Learning" at the Natural Language Processing and Chinese Computing (NLPCC) conference in 2014.

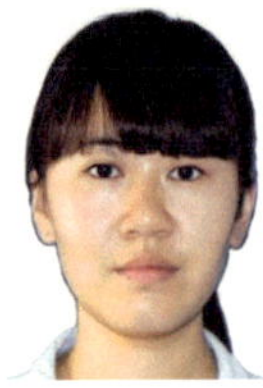

Xuran Pan is currently a lecturer in the College of Artificial Intelligence, Tianjin University of Science and Technology, Tianjin, China. She received a Ph.D. in Electronic Science and Technology from Hebei University of Technology, Tianjin, China, in 2020. Her research interests include image processing, machine learning, deep learning, and their applications in earth vision.

# Contents

# Preface

Nowadays, deep learning and reinforcement learning have become some of the hottest research directions in computer science. They can solve complex problems such as natural language processing, computer vision, medical image analysis, and more by training powerful neural networks. The deep learning algorithm has become one of the most important and promising technologies in the field of artificial intelligence. In addition, reinforcement learning can autonomously learn and adjust to maximize rewards, which is expected to solve complex sequential decision tasks, such as intelligent games and robot control.

In recent years, the rapid development and widespread application of deep learning and reinforcement learning have created enormous commercial and social value. This book introduces the latest advances in the fields of deep learning and reinforcement learning, covering a variety of key areas like natural language processing, medicine analysis, and Internet of Things (IoT) device recognition.

This book consists of two sections: "Theory and Algorithms of Deep Learning and Reinforcement Learning" and "Applications of Deep Learning and Reinforcement Learning." Sections I and II contain two and four chapters, respectively. Section I discusses new network structures and algorithms for deep learning and reinforcement learning. Section II explores new deep learning and reinforcement learning solutions to the challenges faced by the fields of natural language processing, medicine analysis, and IoT device recognition.

I would like to express my sincerest gratitude to the editors, authors, and reviewers who have contributed to this book.

Thank you!

**Jucheng Yang, Yarui Chen, Tingting Zhao, Yuan Wang and Xuran Pan**
College of Artificial Intelligence,
Tianjin University of Science and Technology,
Tianjin, China

Section 1

# Theory and Algorithms of Deep Learning and Reinforcement Learning

Chapter 1

# Utilized System Model Using Channel State Information Network with Gated Recurrent Units (CsiNet-GRUs)

*Hany Helmy, Sherif El Diasty and Hazem Shatila*


## Abstract

MIMO: multiple-input multiple-output technology uses multiple antennas to use reflected signals to provide channel robustness and throughput gains. It is advantageous in several applications like cellular systems, and users are distributed over a wide coverage area in various applications such as mobile systems, improving channel state information (CSI) processing efficiency in massive MIMO systems. This chapter proposes two channel-based deep learning methods to enhance the performance in a massive MIMO system and compares our proposed technique to the previous methods. The proposed technique is based on the channel state information network combined with the gated recurrent unit's technique CsiNet-GRUs, which increases recovery efficiency. Besides, a fair balance between compression ratio (CR) and complexity is given using correlation time in training samples. The simulation results show that the proposed CsiNet-GRUs technique fulfills performance improvement compared with the existing literature techniques, namely CS-based methods Conv-LSTM CsiNet, LASSO, Tval3, and CsiNet.


**Keywords:** massive MIMO, FDD, compressed sensing, deep learning, conventional neural network

## 1. Introduction

For fifth-generation wireless communication systems, the massive multiple-input multiple-output (MIMO) system is recognized as a powerful technology.

Such a system can significantly reduce multi-user interference and offer a multi-fold boost in cell throughput by outfitting a base station (BS) with hundreds or even thousands of antennas in a dispersed or centralized way. Utilizing channel state information (CSI) at base stations is the primary method for obtaining this potential benefit (BSs). The downlink channel state information (CSI) in modern frequency division duplex (FDD) MIMO systems (such as long-term evolution Release-8) is collected at the user equipment throughout the training phase and transmitted back to the BS via feedback links.

To minimize feedback overhead, vector quantization or codeword-based techniques are frequently used. The feedback quantities generated from these methods are not permitted in a massive MIMO system since they must be scaled linearly with the number of transmit antennas. As shown in [1], the difficulty of CSI feedback in massive MIMO systems has inspired several studies. By using the spatial and temporal correlation of channel state information (CSI), which describes how a signal travels from the transmitter to the receiver and represents the combined effect of, for example, scattering, fading, and power decay with distance, these works have primarily concentrated on reducing feedback overhead. To minimize feedback overhead, vector quantization or code-word-based techniques are frequently used. To minimize feedback overhead, vector quantization or codeword-based techniques are frequently used. The feedback quantities generated from these methods are not permitted in a massive MIMO system since they must be scaled linearly with the number of transmit antennas. As shown in [1], the difficulty of CSI feedback in massive MIMO systems has inspired several studies. By using the spatial and temporal correlation of channel state information (CSI), which describes how a signal travels from the transmitter to the receiver and represents the combined effect of, for example, scattering, fading, and power decay with distance, these works have primarily concentrated on reducing feedback overhead. To minimize feedback overhead, vector quantization or codeword-based techniques are frequently used.

A difficult issue in wireless communications systems is channel estimate during auto-encoding. Most of the time sent signals are reflected and scattered as they reach the receiver. The channel moves over time as a result of the mobility of the transmitter, receiver, or scattering objects. Deep learning (DL) trains massive, multilayered neural networks using lots of training data to approximate how the human brain does a particular activity. Channel State Information Networks (CsiNet), which we created as CSI sensing (or encoder) and recovery (or decoder) networks, include the features listed below in the auto-encoder system (**Figure 1**).

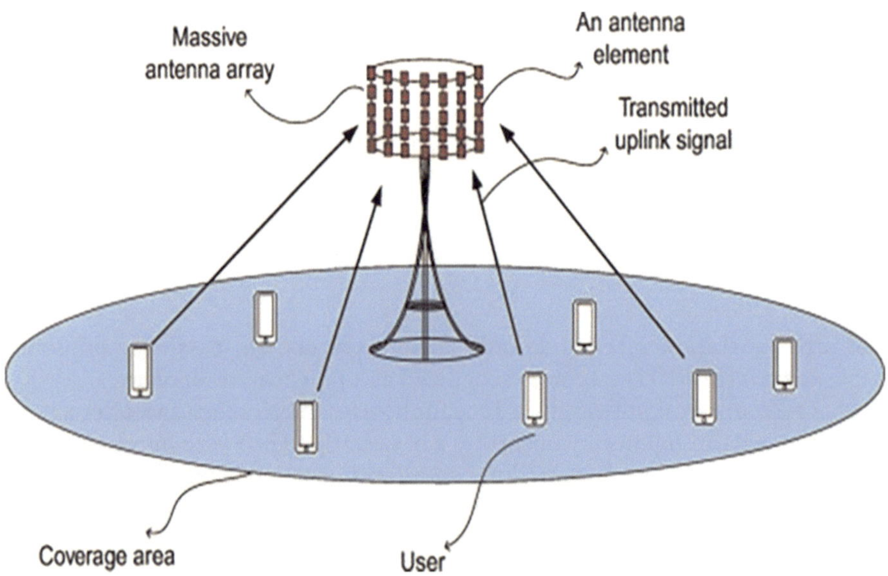


**Figure 1.**
*Enhanced multiple-access for mmWave massive MIMO [2].*

- Encoder: CsiNet learns transformation from original channel matrices to compress representations (codewords) through training data.

- Decoder: CsiNet learns inverse transformation from codewords to original channels; The inverse transformation is not iterative and multiple orders of magnitude faster than iterative algorithms. The algorithm is agnostic to human knowledge of channel distribution and instead directly learns to use the channel structure from training data effectively.

User equipment encodes channel matrices into codewords using the encoder; after the codewords are returned to the BS, it uses the decoder to reconstruct the original channel matrices. The technique can be applied as a feedback protocol in FDD MIMO systems. The autoencoder [3] in deep learning, which is used to learn an encoding for a set of data types for dimensionality reduction, and CsiNet are closely related. To recreate accurate models from CS data, several deep learning (DL) architectures have recently been designed and introduced in [4–6].

DL shows state-of-the-art performance in natural-image reconstruction, but because wireless channel reconstruction is more difficult than image reconstruction, it can also demonstrate that this capability is unclear. The DL-based CSI reduction and recovery strategy is introduced in the current work. The most significant research appears to be [7], in which a closed-loop MIMO system implements DL-based CSI encoding. It differs from previous research that did not consider CSI recovery by demonstrating that, as compared to current CS-based methods, CSI can be recovered with a significantly increased reconstruction quality by DL.

## 2. The structure of channel state information network (CsiNet)

The structure of CsiNet [8] according to Depth wise Separable Convolution in feature recovery reconstruction illustrated in detail, CsiNet remarkably outperforms the CS-based methods. Introducing the CSI network feedback process, which considers a single-cell FDD massive MIMO-OFDM framework, where there is $\mathbf{N_t}$ ($\gg$1) transmit antennas at the BS and a single receiver antenna at the UE, OFDM is with $\mathbf{N_c}$ subcarriers the received signal at the $\mathbf{n^{th}}$ subcarrier can be communicated as:

$$y_n = \tilde{h}_n^H v_n x_n + z_n \tag{1}$$

where $\tilde{\boldsymbol{h}}_{\boldsymbol{n}}^{\boldsymbol{H}}$ and $\boldsymbol{y_n} \epsilon\ \mathbb{C}\ \mathbf{N_t\ x\ 1}$ is the channel frequency response vector and the precoding vector at the $n^{th}$ subcarrier, separately, $\boldsymbol{x_n}$ represents the transmitted information image, $\boldsymbol{z_n}$ is the additive noise or obstruction and $(\cdot)^{\mathrm{H}}$ is a conjugate transpose. In the FDD system, improving feedback links through UE and BS, focus on the feedback scheme which allows autoencoder processing, assume:

$\hat{\mathbf{H}} = \left[\tilde{\mathbf{h}}_1 \ldots \tilde{\mathbf{h}}_{Nc}\right]^H \epsilon\ \mathbb{C}^{Nc\ x\ Nt}$ in CSI stacked in the spatial frequency domain, which means the UE should return $\hat{\mathbf{H}}$ to the BS through feedback links, and in the feedback system, the total number parameter is $\mathbf{N_t N_c}$, using a 2D (DFT) discrete Fourier transform, which introducing $\tilde{\mathbf{H}}$ can be improved in the angular-delay domain to reduce feedback overhead:

$$H = F_d \cdot \tilde{H}F_a^H \quad (2)$$

where $\mathbf{F_d}$ **and** $F_a$ **are** $\mathbf{N_c}$ **X** $\mathbf{N_c}$ **and** $\mathbf{N_t}$**X** $\mathbf{N_t}$ DFT matrices, respectively. So, considering the COST 2100 as was illustrated in [9] channel model as shown in **Figure 2**. depending on a uniform linear array (ULA), **H** has a small fraction of significant components. According to the delay domain, the first $\mathbf{N_c}$ rows of **H** contain values, retain the first $\mathbf{N_c}$ Rows of **H** and remove remaining rows. In a massive MIMO system, the total number of feedback parameters can be reduced to **N** = $\mathbf{N_c}$ $\mathbf{N_{t.}}$ So, we design the encoder *S*,

$$S = f_{en}(H) \quad (3)$$

We can convert **H** into a codeword M vector, where **M < N,** and design the decoder inverse transformation from the codeword to H original channel.

$$H = f_{de}(S) \quad (4)$$

The European Cooperation in Science and Technology COST 2100 channel model is a GSCM that can reproduce the stochastic properties of massive MIMO channels

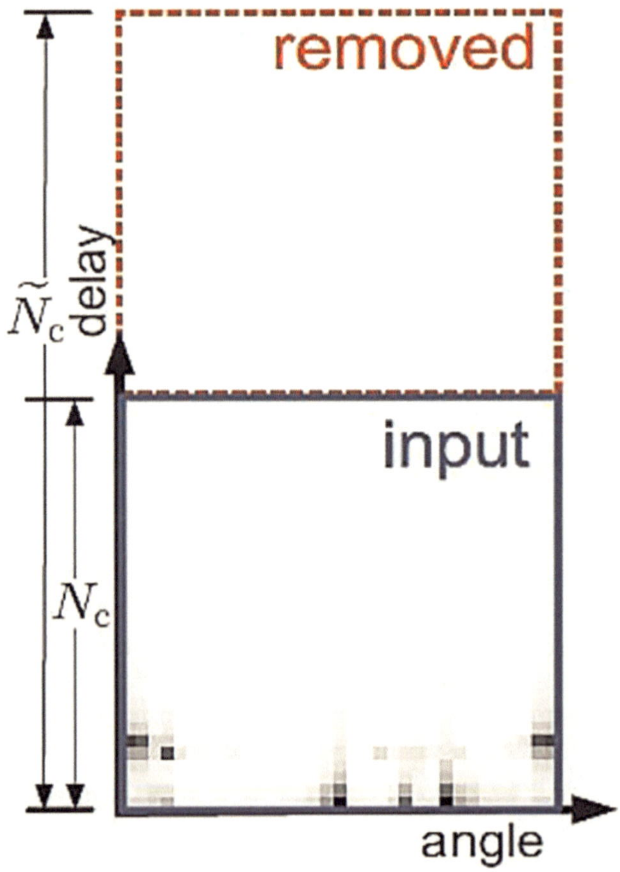


**Figure 2.**
*A plot of the strength of H ϵ* $\mathbb{C}^{32x32}$ *[8].*

over time, frequency, and space. A multi-path component MPC is characterized in delay and angular domains by its delay, angle of departure (Azimuth of Departure (AoD), Elevation of Departure (EoD), and angle of arrival (Azimuth of Arrival (AoA), Elevation of Arrival (EoA)). The MPCs with similar delays and angles are grouped into multi-path clusters. The MATLAB implementation of C2CM supports both single-link and multiple-link MIMO channel access indoor (285 MHz) and semi-urban (5.3 GHz) channel scenarios. An overview of the C2CM is presented in a detailed description of the channel model. The parameterization of the C2CM in indoor scenarios is detailed while discussing semi-urban scenarios. On the other hand, it gives the massive multiple-input multiple-output MIMO extensions of the C2CM; The C2CM is implemented in MATLAB, while the semi-urban channel scenario is implemented in [9]. Furthermore, the MATLAB implementation of C2CM with massive MIMO extensions. However, the data generated in these MATLAB implementations are not presented as potential datasets to validate multi-path clustering methods and even in the well-known clustering approaches.

## 3. Recurrent unit system model

### 3.1 The structure

It can adaptively capture capable of adaptively capturing dependencies from lengthy data sequences without removing data from previous stages of the sequence due to the gated recurrent unit structure [10]. This is accomplished by its gating units, which are related to those in long short-term memory LSTMs, and which resolve the vanishing/exploding gradient problem of conventional RNNs. These gates control the information that should be retained or discarded at each time step. The GRU operates like an RNN, except for its internal gating mechanisms, where sequential input data is absorbed by the GRU cell at each time step along with memory, also known as the hidden state [11]. The RNN cell and the following input data in the sequence are then fed with the hidden state once more (**Figure 3**).

Fully gated unit

Initially, for $t = 0$, the output vector is $h_0 = 0$.

$$z_t = \rho_g \left(W_z x_t + U_Z h_{t-1} + b_z\right) \tag{5}$$

$$r_t = \rho_g \left(W_r x_t + U_r h_{t-1} + b_r\right) \tag{6}$$

$$h_t = z_t \odot h_{t-1} + (1 - z_t) \odot \phi_h \left(W_h x_t + U_h \left(r_t \odot h_{t-1}\right) + b_n\right) \tag{7}$$

Were, $x_t$ input vector, $h_t$ output vector, $z_t$ update gate vector, $r_t$ reset gate vector and $W$, $U$, and $b$ denote matrices and vectors, respectively.

**Activation functions:** $-\boldsymbol{\rho_g}$ Original sigmoid activation, $\boldsymbol{\phi_h}$ For the initial hyperbolic tangent, Alternative activation functions can be used, provided the $\rho_g(x) \in [0,1]$.

It is possible to construct alternative forms by modifying $z_t$ and $r_t$.

GRU's ability to hold on to long-term dependencies or memory stems from the gated recurrent unit cell's computations to produce the hidden state. At the same time, LSTMs have two different states passed between the cell state and hidden state, which carry the long and short-term memory, respectively GRUs only have one hidden state transferred between time steps. This hidden state can hold both long-term and short-

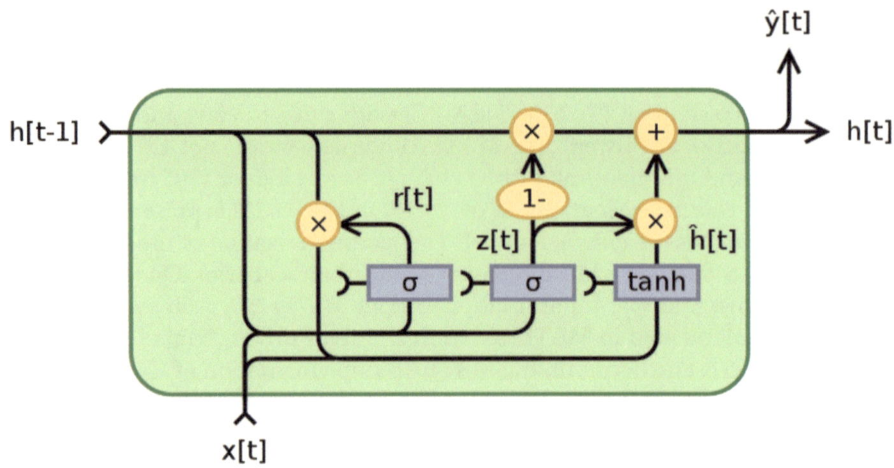


**Figure 3.**
*Gated recurrent unit, fully gated version [12].*

term dependencies at the same time due to the gating mechanisms and computations that the hidden state and input data go through.

The GRU cell contains only two gates: The Update gate and the Reset gate; like the gates in LSTMs, the GRU gates are trained to selectively filter out any irrelevant information while keeping what's useful. These gates are essentially vectors containing values between 0 and 1, multiplying with the input data or hidden state.

A zero (0) value in the gate vectors indicates that the input or hidden state's corresponding data is unimportant and will, therefore, return as a zero.

On the other hand, a one (1) value in the gate vector means that the corresponding data is essential and will be used. Reset gate: In the first step, we'll create the Reset gate; this gate is derived and calculated using both the hidden state from the previous time step and the input data at the current time step (**Figure 4**).

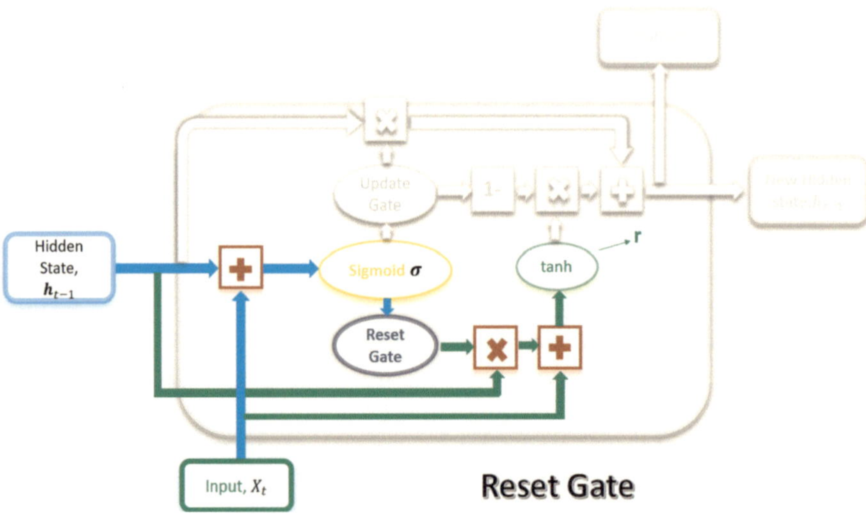


**Figure 4.**
*Reset gate flow [13].*

Mathematically, this is achieved by multiplying the previous hidden state and current input with their respective weights, summing before passing the sum through the sigmoid function.

The sigmoid function will transform the values to fall between 0 and 1, allowing the gate to filter between the less-important and more-important information in the subsequent steps.

$$Gate_{reset} = \sigma \left( W_{inputreset} \cdot x_t + W_{hiddenreset} \cdot h_{t-1} \right) \tag{8}$$

When the entire network is trained through back-propagation, the weights in the equation will be updated such that the vector will learn to retain only the valuable features. The previous hidden state will first be multiplied by a trainable weight and will then undergo an element-wise multiplication Hadamard product with the reset vector. This operation will decide which information will be kept from the previous time steps and the new inputs.

Simultaneously, the current input will also be multiplied by a trainable weight before being summed with the reset vector's product and the previous hidden state above. Finally, a non-linear activation tanh function will be applied to the result to obtain r in the equation below.

$$\mathrm{r} = \tanh \left( gate_{reset} \odot (W_{h1} \cdot h_{t-1}) + W_{x1} \cdot x_t \right) \tag{9}$$

Update gate: next, we'll create the Update gate, like the Reset gate; the gate is computed using the previous hidden state and current input data (**Figure 5**). Both the Update and Reset gate vectors are created using the same formula, but the weights multiplied with the input and hidden state are unique to each gate, which means that each gate's final vectors are different; This allows the gates to serve their specific purposes.

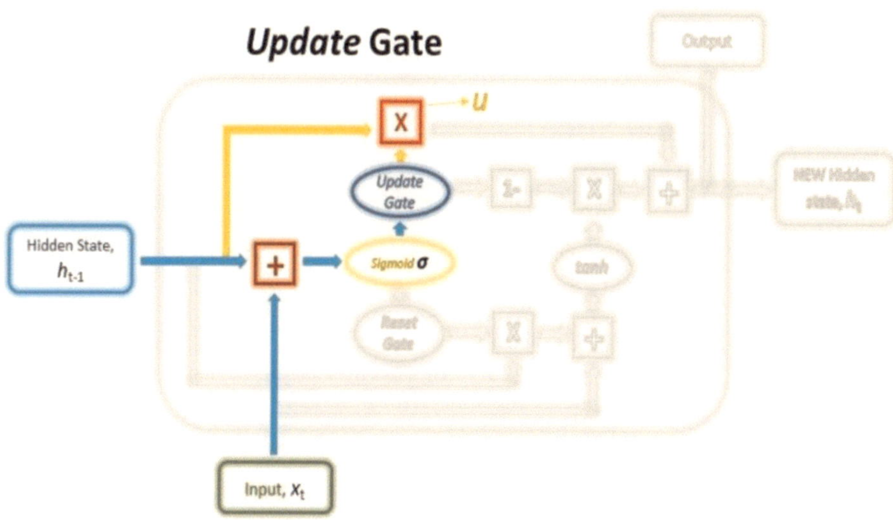


**Figure 5.**
*Update gate flow [13].*

$$gate_{update} = \sigma\left(W_{inputupdate} \cdot x_t + W_{hiddenupdate} \cdot h_{t-1}\right) \tag{10}$$

The Update vector will undergo element-wise multiplication with the previous hidden state to obtain u in our equation below, which will be used to compute our final output later.

$$\mathrm{u} = gate_{update} \odot h_{t-1} \tag{11}$$

The Update vector will also be used in another operation later when obtaining our final output.

The purpose of the Update gate here is to help the model determine how much of the past information stored in the previous hidden state needs to be retained for the future. Combining the outputs: In the last step, we will be reusing the Update gate and obtaining the updated hidden state (**Figure 6**).

This time, we will be taking the element-wise inverse version of the same Update vector (1—Update gate) and doing an element-wise multiplication with our output from the Reset gate, r. This operation's purpose is for the Update gate to determine what portion of the new information should be stored in the hidden state. Lastly, the result of the above operations will be summed with our output from the Update gate in the previous step, u.

This will give us our new and updated hidden state; We can use this new hidden state as our output for that time step by passing it through a linear activation layer.

$$h_t = \mathrm{r} \odot \left(1 - gate_{update}\right) + \mathrm{u} \tag{12}$$

The Reset gate determines which parts of the previous hidden state are to be combined with the current input to propose a new hidden state, and the Update gate determines how much of the previous hidden state is to be retained and what part of the new proposed hidden state derived from the Reset gate is to be added to the final

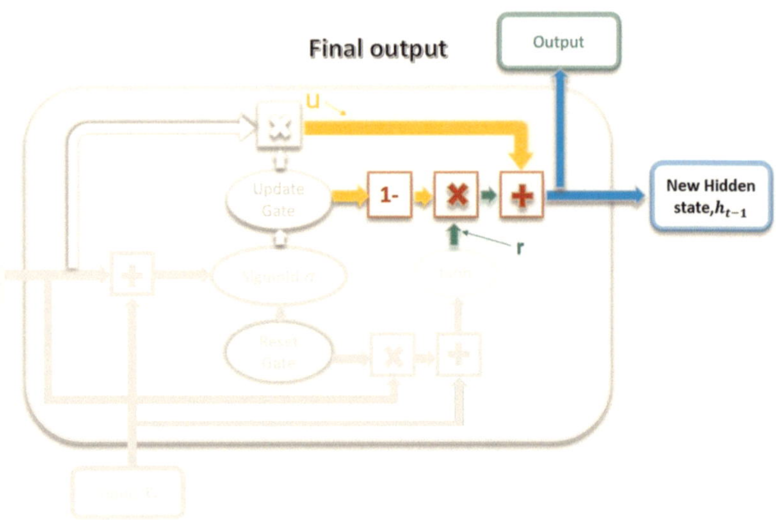


**Figure 6.**
*Final output computations [13].*

hidden state. This solves the Vanishing/Exploding Gradient Problem. The network chooses which components of the previous hidden state to keep in memory while discarding the rest when the Update gate is first multiplied with it. When it uses the Reset gate's inverse gate to filter the proposed new hidden state from the Update gate, it then fills in the gaps in the information that were previously missing. The network can maintain long-term dependencies as a result. If the Update vector values are close to 1, the Update gate may decide to keep most of the previous memories in the hidden state rather than recalculating or altering the hidden state entirely.

When training a recurrent neural network RNN, the vanishing or exploding gradient problem can happen, especially if the RNN is processing lengthy sequences or has multiple layers. The network's weight is updated in the right direction and by the right amount using the error gradient that was calculated during training. However, this gradient is determined using the chain rule, beginning at the end of the network. As a result, for long sequences, the gradients will undergo continuous matrix multiplications and either disappear (vanish) or explode (explode) exponentially.

A gradient that is too small will prevent the model from effectively updating its weights, whereas a gradient that is too large will make the model unstable.

Due to the addictive nature of the Update gates, the long short-term memory (LSTM) and gated recurrent units (GRUs) can keep most of the existing hidden state while adding new content on top of it, unlike traditional RNNs that always replace the entire hidden state content at each time step.

This prevents the additional operations from causing the error gradients to vanish or explode too quickly during backpropagation. Utilizing alternative activation functions, like ReLU, which does not result in a small derivative, is the simplest solution. Another option is residual networks, which offer residual connections directly to earlier layers. In a feedforward network (FFN), the backpropagated error signal typically decreases (or increases) exponentially as a function of the distance from the final layer. This technique is referred to as the vanishing gradient.

## 4. Design of the CsiNet-GRUs system model

We enhance the architecture in this chapter by considering time correlation. The recurrent convolutional architecture that has been used to represent and reconstruct images successfully provides references for our work. The basic idea is to extract spatial features and inter-frame correlation using convolutional neural networks (CNN) and recurrent neural networks (RNN), respectively. The following is a summary of our contribution to this chapter. CsiNet is extended with a gated recurrent unit network, a common type of recurrent neural network, and a DL-based CSI feedback protocol for FDD MIMO systems is proposed (RNN). CsiNet-GRUs is a proposed network that modifies the CNN-based convolutional neural network CsiNet for channel state information compression and initial recovery and uses a gated recurrent unit technique to extract time correlation for further resolution improvement.

CsiNet-GRUs exhibit remarkable robustness to compression ratio (CR) reduction and enable real-time and extensible channel state information (CSI) feedback applications without considerably increasing overhead compared with CsiNet, to reduce feedback overhead, we can exploit the following observations: **Observation 1 (angular-delay domain sparsity):** $\boldsymbol{H}_t$ can be transformed into an approximately sparsified matrix $H_t'$ in the angular-delay domain via 2D discrete Fourier transform (2D-DFT) by $H_t' = F_d H_t F_a$, where $F_d \in \mathbb{C}^{N_c \times N_c}$ and $F_a \in \mathbb{C}^{N_t \times N_t}$ are two DFT matrices. First, due to limited multipath time delay, performing DFT on frequency domain channel vectors (i.e., column vectors of $H_t$) can transform $H_t$ into a sparse matrix in the delay domain, with only the first $N_c'$ ($<N_c$) rows have distinct non-zero values. Secondly, the channel matrix is sparse in a defined angle domain by performing DFT on spatial domain channel vectors (i.e., row vectors of $H_t$) If the number of transmit antennas, $N_t \to +\infty$, is very large. Usually, $H_t'$ is only approximately sparse for finite $N_t$ which challenges conventional compressed sensing methods. Therefore, we will propose a DL-based feedback architecture without sparsity prior constraint. We perform sparsity transformation to decrease parameter overhead and training complexity.

We retain the first $N_c'$ non-zero rows and truncate $H_t'$ to a $N_c' \times N_t$ matrix, $H_t''$, which reduces the total number of parameters for feedback to $N = N_c' N_t$.

**Observation 2 (correlation within coherence time):** The user equipment motion during communication results in a Doppler spread, time-varying characteristics of wireless channels with the maximum movement speed denoted as **v**, coherence time can be calculated as:

$$\Delta t = \frac{C}{2\, v\, fo} \tag{13}$$

where $f_0$ is the carrier frequency, and **c** is the speed of light. The CSI within $\mathbf{\Delta t}$ is considered correlated with one other. Therefore, instead of independently recovering CSI, the BS can combine the feedback and previous channel information to enhance the subsequent reconstruction.

We set the feedback time interval as $\boldsymbol{\delta t}$ and place **T** adjacent instantaneous angular-delay domain channel matrices into a channel group, i.e.,

$$\{H_t''\}_{t=1}^{T} = \{H_1'', \dots H_t'', \dots, H_T''\} \tag{14}$$

The group exhibits correlation property if **T** satisfies $\mathbf{0} \leq \boldsymbol{\delta t} \cdot \mathbf{T} \leq \mathbf{\Delta t}$.

In this chapter, we design an encoder, $S_t = f_{en}(H_t'')$, at the UE to compress each complex-valued $H_t''$ of $\{H_t''\}_{t=1}^{T}$ into an M-dimensional real-valued codeword vector $S_t$ ($\mathbf{M} < \mathbf{N}$). If two real number matrices are used to represent the real and imaginary parts of $H_t''$, then CR will be $M/2N$.

We also design a decoder with a memory that can extract time correlation from the previously recovered channel matrices, $\hat{H}_1'', \dots, \hat{H}_{t-1}''$ and combine them with the received $S_t$ for the current reconstruction channel, $\hat{H}_t'' = f_{de}(S_t; \hat{H}_1'', \dots, \hat{H}_{t-1}'')$, where $\mathbf{1} \leq \mathbf{t} \leq \mathbf{T}$. Then, inverse 2D-DFT is performed to obtain the original spatial frequency channel matrix; the Channel state information network demonstrates remarkable performance in CSI sensing and reconstruction. However, the resolution degrades at low CR because it only focuses on angular-delay domain sparsity (Observation 1) and ignores the time correlation (Observation 2) of time-varying massive MIMO channels, the two observations are like the spatial and inter-frame correlations of videos.

Motivated by RCNN, which excels in extracting spatial-temporal features for video representation, we will extend CsiNet with GRU to improve CR and recovery quality

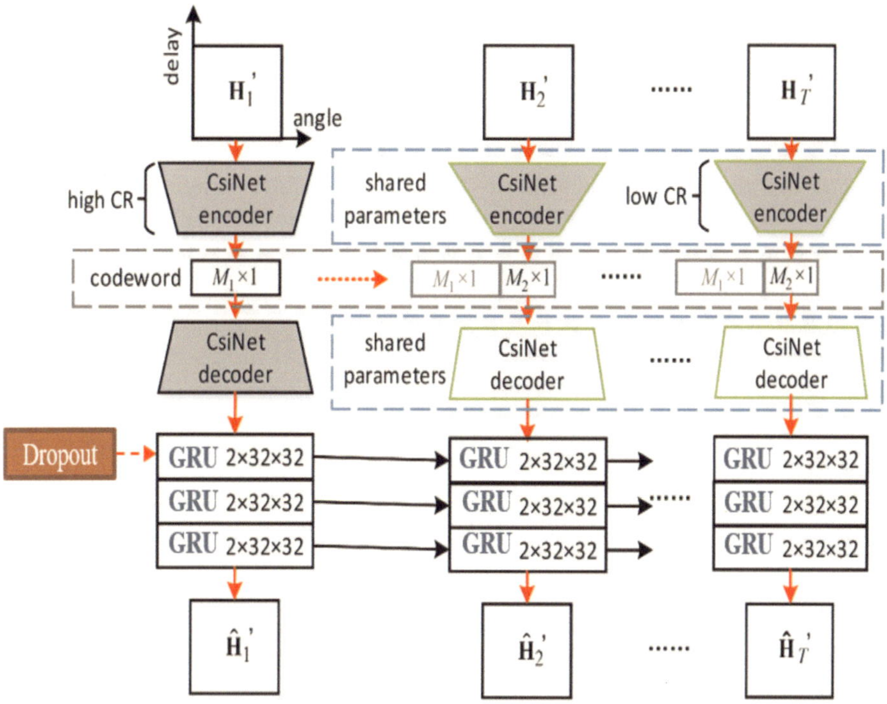


**Figure 7.**
*The structure of the proposed CsiNet-GRUs using dropout technique [14].*

trade-off. We will also introduce the multi-CR strategy to implement variable CRs on different channel matrices; The proposed CsiNet-GRU is illustrated in **Figure** 7. with CsiNet. Our model includes the following two steps: angular-delay domain feature extraction, correlation representation, and final reconstruction. Each GRU has an inherent memory unit that, for future prediction, can hold previously extracted information for a long time. A $3 \times 3$ convolutional layers and an M-unit dense layer for sensing, and a dense layer with $\mathbf{2N_c'N_t}$ units should be considered to facilitate comparison with the results of the CsiNet structure given in [8] and two decoders from RefineNet for reconstruction as shown in **Figure** 7, each RefineNet comprises channel into four $3 \times 3$ convolutional layers with different channel sizes.

The CsiNet decoder's output generates a sequence, and the length of every sequence is **T**, which is then fed into a three-layer GRU. All low-CR CsiNet's shown in **Figure** 7. share the same network parameters, i.e., weights and bias, because they perform the same work, which dramatically reduces parameter overhead. Furthermore, the architecture can be easily rescaled to perform on channel groups with different **T** if the value of **T** changes to adapt to the channel-changing speed and feedback frequency; A low-CR CsiNet will be reused $(T - 1)$ time instead of making $(T - 1)$ copies in practice. The gray blocks in **Figure** 7 load parameters from the original CsiNet's as pre-training before end-to-end training with the entire architecture. This method can alleviate vanishing gradient problems due to long paths from CsiNet's to GRUs. We use GRUs to extend the CsiNet decoders for time correlation extraction and final reconstruction. Gated recurrent units have inherent memory cells and can keep the previously extracted information for a long period for later prediction. In particular, the CsiNet decoders' outputs form a sequence of length **T** before being fed into three-layer GRUs. Each GRU has a $\mathbf{2N'_cN_t}$; The hidden unit is the same as the size of the output. Then the final output is reshaped into two matrices as the final recovered $\hat{\boldsymbol{H}}''_t$; This allows the CR-CsiNet encoder to send to the rest $\mathbf{T-1}$.

Because less information is required due to channel correlation, the channel matrix performs operations, $\boldsymbol{M_2} \times \mathbf{1}$ codewords $(\boldsymbol{M_1} > \boldsymbol{M_2})$, are generated. The $(\boldsymbol{T}-\mathbf{1})$ codewords are all concatenated with the first codeword $\boldsymbol{M_1} \times \mathbf{1}$ before being fed into the low-CR CsiNet decoder to utilize feedback information fully. Each CsiNet outputs two matrices with size $(\boldsymbol{N'_c} \times \boldsymbol{N_t})$ as extracted features from the angular delay domain as the final recovered $\hat{\mathbf{H}}''_t$. The spatial frequency domain CSI can then be obtained via inverse 2D-DFT. At each time step, the GRUs implicitly learns time correlation from the previous inputs and then merge them with the current inputs to increase low CR recovery quality.

## 4.1 The dropout technique

Dropout: During training, randomly selected neurons are ignored and "dropped out." This means that their contribution to downstream neuron activation is removed temporally on the forward pass, and no weight updates are applied to the neuron on the backward pass. Dropout can be implemented on any hidden layer in the network; the visible or input layer, as well as the term "dropout," refers to dropping out units (hidden and visible) in a neural network. Dropout is a regularization method used when training the network, as illustrated in **Figure 8**. It is possible that the input and loop connections to the gated recurrent unit (GRU) in **Figure** 7 are not excluded from activation and weight updates. Depending on the framework, the dropout regularization Training Phase: Ignore (zero out) a random fraction, p, of nodes for each hidden layer, training sample, and iteration (and corresponding activations). A phase of testing: Use all activations but reduce them by a factor of p. (to account for the missing activations during training). Dropout is a regularization method used when training the network, as shown in **Figure 8**. However, it does not always exclude the input and loop connections to the gated recurrent unit (GRU) from activation and weight updates, as shown in **Figure** 7. To reduce overfitting and improve the efficiency of the CsiNet-GRU structure, a neural network approach is used. We stated

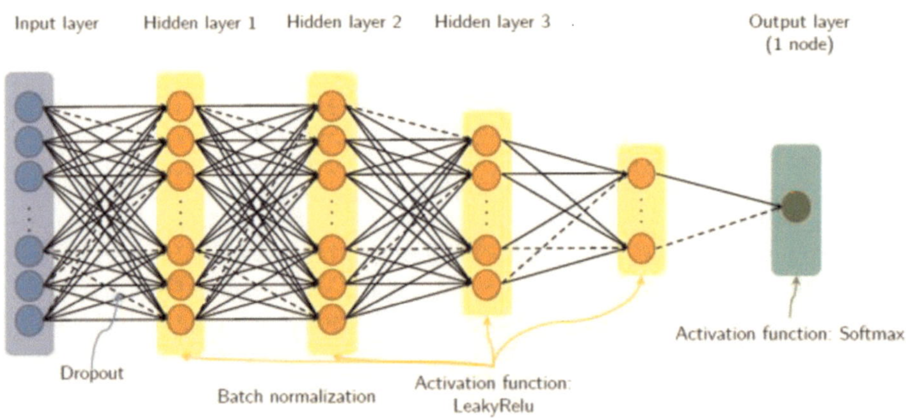


**Figure 8.**
*Neural network with dropout architecture [15].*

that the effect on "downstream neurons" activation during the forward process would be temporarily removed and that no weight update for "backward propagation to neurons" would be applied [15].

Training Phase: Ignore (zero out) a random fraction, p, of nodes for each hidden layer, training sample, and iteration (and corresponding activations). A phase of testing: Use all activations but reduce them by a factor of p. (to account for the missing activations during training). Dropout is a regularization method used when training the network, as shown in **Figure 8**. However, it does not always exclude the input and loop connections to the gated recurrent unit (GRU) from activation and weight updates, as shown in **Figure 7**. Depending on the framework, the dropout regularization approach used in neural networks is used to reduce overfitting and improve the efficiency of the CsiNet-GRU structure. We stated that the effect on "downstream neurons" activation during the forward process would be temporarily removed and that no "backward propagation to neurons" weight update would be applied [15]. Training Phase: Ignore (zero out) a random fraction, p, of nodes for each hidden layer, training sample, and iteration (and corresponding activations). A phase of testing: Use all activations but reduce them by a factor of p. (to account for the missing activations during training).

Some observations: Dropout forces a neural network to learn more robust features that can be used in conjunction with the random subsets of many other neurons. Dropout roughly doubles the number of iterations needed to converge; however, each epoch's training time is less, and during the testing phase, the entire network is considered, and each activation is reduced by a factor of p. When training the network in the proposed structure, the input and recurrent connections to the GRU unit may not be excluded from activation and weight updates.

There are two dropout parameters in RNN layers: *dropout*, applied to the first operation on the inputs, and *recurrent dropout* applied to the other operation on the recurrent inputs. It is worth mentioning that interested in designing the encoder which can transform the channel matrix into an M-dimensional vector (codeword), where $\mathbf{M} < \mathbf{N}$. Thus, define the data compression ratio $\boldsymbol{\gamma}$ as $(\gamma = \boldsymbol{M}/2\boldsymbol{N_t}\boldsymbol{N_c})$.

The encoder first extracts CSI features via a convolutional layer with two $3 \times 3$ filters, followed by an activation layer. A fully connected (FC) layer with M neurons is then used to compress the CSI features to lower dimensions. The compression ratio (CR) of this encoder can be expressed as $\boldsymbol{CR} = \mathbf{1}/\boldsymbol{\gamma}$. The final reconstruction of the CSI is performed by three $2\boldsymbol{N}_c'\boldsymbol{N}_t$ unit GRUs with dropout techniques.

Moreover, adopting depth-wise separable convolutions in feature recovery reduces the model's size and interacts with information between channels and introducing the delay $\boldsymbol{\theta}$ as a parameter used in the encoder and decoder, i.e., $\boldsymbol{\theta} = \{\boldsymbol{\theta_{en}}, \boldsymbol{\theta_{de}}\}$.It is worth mentioning that $\boldsymbol{H}_t''$ are standardized with all components scaled into the [0; 1], and this standardization is required for CsiNet.

## 5. COST 2100 data sets and parameters

The COST 2100 channel model is a geometry-based stochastic channel model (GSCM) capable of reproducing the stochastic properties of multi-link multiple-input multiple-output channels across time, frequency, and space. As a result, there is no doubt that more advanced channel estimation methods and good measurement campaigns for parameterization and validation are required for the successful development and long-term use of the COST 2100 channel model. Multiple-input multiple-output (MIMO) is a technology that enables faster and more reliable transmissions over wireless channels.

The COST 2100 model simulates MIMO channels and generates training samples; we set the MIMO-OFDM system to work on a 20 MHz bandwidth using a uniform linear array (ULA). The parameters utilized in indoor and outdoor channel scenarios are given in **Table 1**; Data sets are generated by randomly setting different start places for indoor and outdoor scenarios and performing the simulations at CR values with the first channel $\boldsymbol{H}_1''$ they were compressed under 1/4. **Table 1** shows the training, validation, and testing sets; some parameters are preloaded from the CsiNet for initialization (epochs from 500 to 1000, learning rate of 0.001, and batch size of 200), as shown in **Table 1**.

We compare the proposed architecture's performance with previous similar modeling approaches of channel state information (CSI) with different deep learning approaches, namely Conv-LSTM CsiNet, LASSO, TVAL3 [16], and CsiNet, utilizing the default setup in the open-source codes of the previously mentioned techniques for reproduction.

TVAL3 uses a minimum total variation method that provides remarkable recovery quality and high computing efficiency, while LASSO uses simple sparse priors to achieve good performance. In the feature extraction and recovery modules of Convolutional-LSTM CsiNet, RNN, and depth-wise separable convolution were used.

The term "training" refers to the process of determining which parameters to use in a given dataset. We run the modeling CsiNet-GRUs on Collaboratory (python) according to zero configuration required, free access to GPUs, and easy sharing training and testing of the CsiNet, Conv-LSTM CsiNet, and CsiNet-GRUs on python colab editor.

| **MIMO OFDM** | **20 MHz bandwidth** | |
|---|---|---|
| $H$ | $32 \times 32$ | |
| $N_t$ | 32 Antennas | |
| $N_C$ | 1024 Subcarriers | |
| Training samples | 100,000 | |
| Validation samples | 30,000 | |
| Testing samples | 20,000 | |
| Epochs | 500–1000 | |
| Learning rate | 0.001 | |
| Batch size | 200 | |
| $\partial t$ | 0.04 s | |
| $T$ | 10 s–100 s | |
| $CR$ | 4, 16, 32, 64 | |
| **Channel model** | **Indoor scenario** | **Outdoor scenario** |
| Frequency, $f$ | Pico, 5.3 GHz, | Rural, 300 MHz |
| Speed, $v$ | 0.0036 km/h | 4.24 km/h |
| $\Delta t$ | 30 s | 0.56 s |

**Table 1.**
*COST 2100 model DATA-SETS and system, parameters.*

Comparisons are made using the normalized mean square error, cosine similarity, accuracy, and run-time in the indoor and outdoor channels, as well as the complexity factored in. The Normalized Mean Square Error measures and reflects the mean relative scatters.

The normalization of the MSE assures that the metric will not be biased when the model overestimates or underestimates the predictions. So, the normalized mean square error (NMSE) utilized for comparisons quantifies the difference between the input $\{\boldsymbol{H}_t\}_{t=1}^{T}$ and the output $\{\hat{\boldsymbol{H}}_t\}_{t=1}^{T}$ in both proposed techniques CsiNet-GRUs are given by:

$$NMSE = \mathbb{E}\left\{\frac{1}{T}\sum_{t=1}^{T}\left\|H_t'' - \hat{H}_t''\right\|_2^2 / \left\|H_t''\right\|_2^2\right\} \tag{15}$$

The correlation coefficient is a statistical measure of the strength of the relationship between the relative movements of two variables. The values range between −1.0 and 1.0. A correlation of −1.0 shows a perfect negative correlation, while a correlation of 1.0 shows a perfect positive correlation.

To measure the degree of similarity between the actual channel $\boldsymbol{h}_{n,t}$ and the estimated channel value $\widehat{\boldsymbol{h}}_{n,t}$ of the $\boldsymbol{n}^{th}$ subcarrier at time $\boldsymbol{t}$, using the cosine similarity coefficient $\boldsymbol{\rho}$, in CsiNet-GRUs which is given as:

$$\rho = \mathbb{E}\left\{\frac{1}{T}\frac{1}{N_c}\sum_{t=1}^{T}\sum_{n=1}^{N_c}\frac{\left|\widehat{h}_{n,t}^{H} h_{n,t}\right|}{\left\|\widehat{h}_{n,t}\right\|_2 \left\|h_{n,t}\right\|_2}\right\} \tag{16}$$

Where $\widehat{\boldsymbol{h}}_{n,t}$ denotes the reconstructed channel vector of the nth subcarrier at time **t**. **ρ** can measure the quality of the beamforming vector when the vector is set as $\boldsymbol{v}_{n,t} = \widehat{\boldsymbol{h}}_{n,t} / \left\|\widehat{\boldsymbol{h}}_{n,t}\right\|_2$ since the UE will achieve the equivalent channel $\widehat{\boldsymbol{h}}_{n,t}^{H}\boldsymbol{h}_{n,t} / \left\|\widehat{\boldsymbol{h}}_{n,t}\right\|_2$.

Introducing a new parameter for comparison, which calling accuracy defining it as the ratio of the number of correct predictions to the total number of input samples, that means accuracy is the ratio of the recovered channel vector to the original channel vector $\{\boldsymbol{H}_t''\}_{t=1}^{T} / \boldsymbol{H}_1''$ so the accuracy in CsiNet-GRUs is defined as:

$$Accuracy = \mathbb{E}\left\{\frac{1}{T}\frac{1}{N_c}\sum_{t=1}^{T}\sum_{n=1}^{N_c}\frac{\left|\widehat{h}_{n,t}^{H}\right|}{\left\|h_{n,t}\right\|_2}\right\} \tag{17}$$

## 6. Comparison of results with different techniques

**Figures 9** and **10** show the relationship between CR and NMSE for all structures in indoor and outdoor scenarios. **Figure 9** shows that the proposed CsiNet-GRUs have the lowest NMSE, whereas **Figure 10** shows that it has the lowest NMSE among others except for Conv-LSTM CsiNet at CR > 20. **Figures 11** and **12** show the relationship between the CR and accuracy for all structures in indoor and outdoor scenarios.

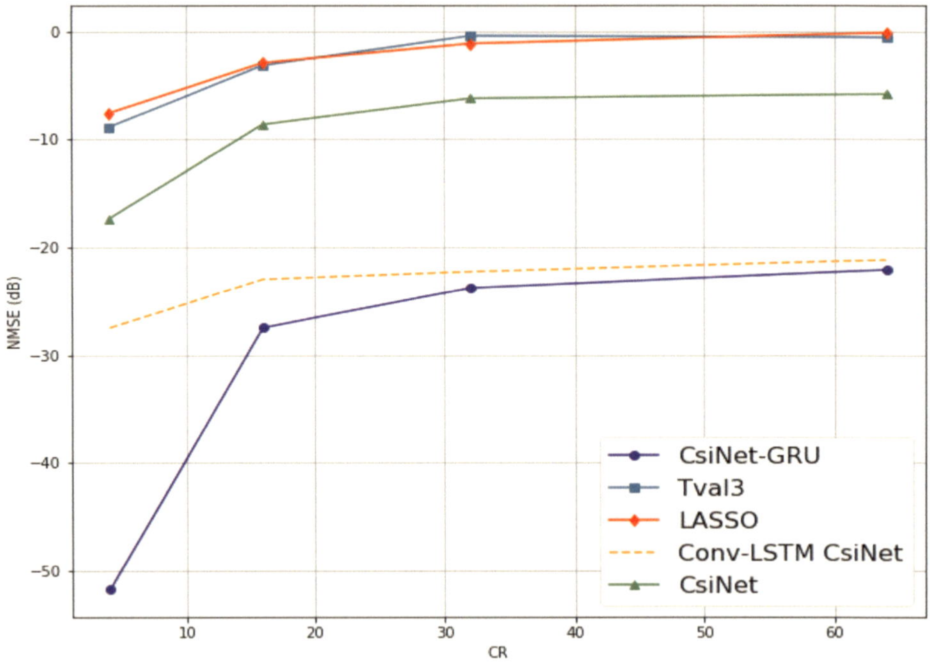


**Figure 9.**
*NMSE (dB) performance comparison between CS methods INDOOR scenario.*

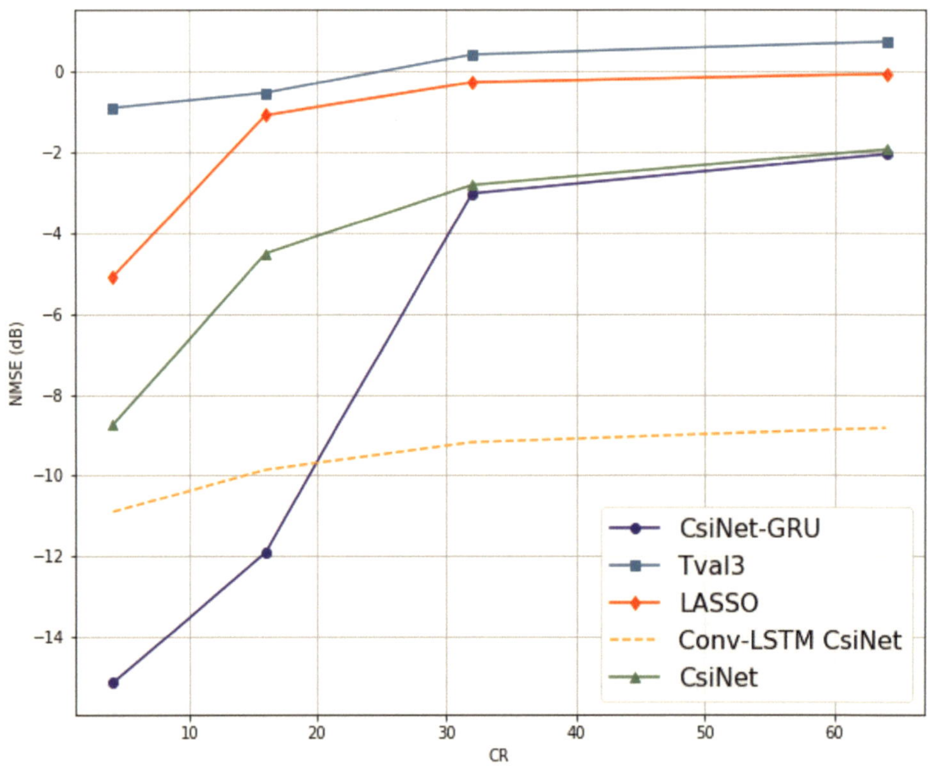


**Figure 10.**
*NMSE (dB) performance comparison between CS methods OUTDOOR scenario.*

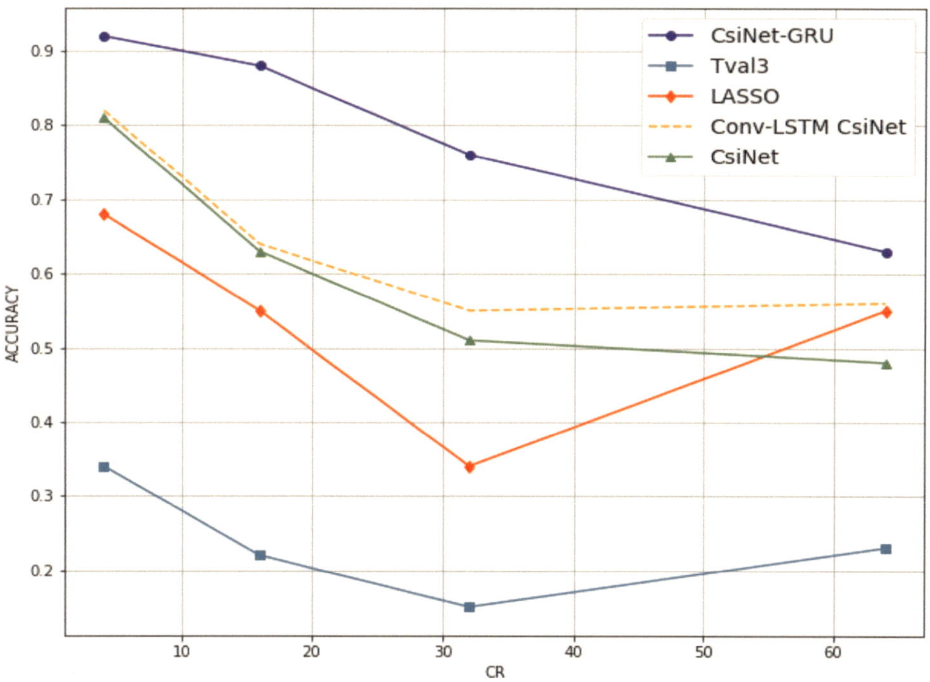


**Figure 11.**
*Accuracy performance comparison between CS methods INDOOR scenario.*

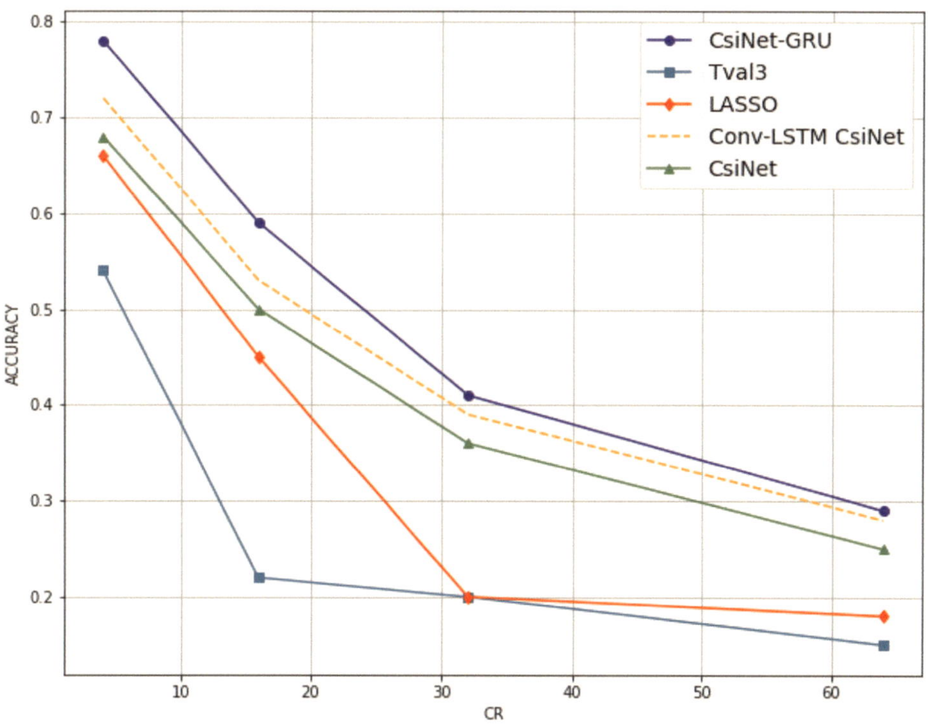


**Figure 12.**
*Accuracy performance comparison between CS methods OUTDOOR scenario.*

The CsiNet-GRUs outperform the other structures, with higher accuracy observed at lower CR values. **Figures 13** and **14** illustrate the relation between the cosine similarity (ρ) and CR in indoor and outdoor scenarios for all structures. Again, the proposed CsiNet-GRUs outperform the other structures, and besides, it exhibits a near-linear performance with the lowest slope.

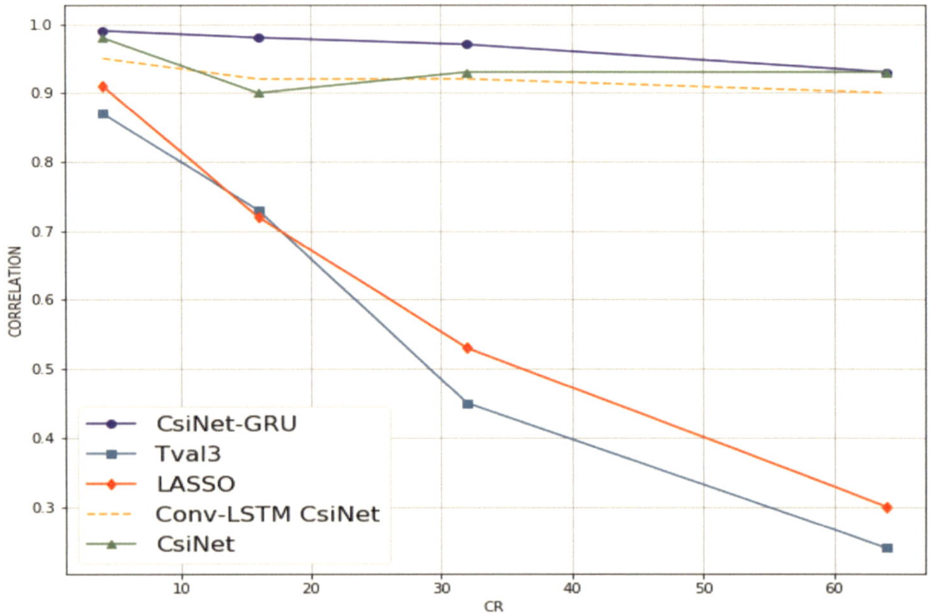


**Figure 13.**
*ρ Performance comparison between CS methods INDOOR scenario.*

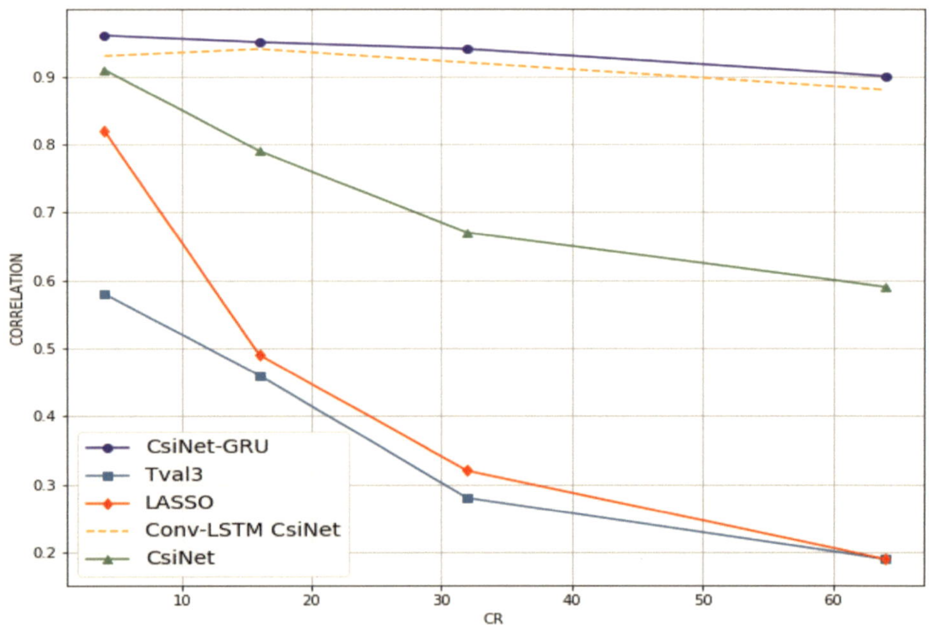


**Figure 14.**
*ρ Performance comparison between CS methods OUTDOOR scenario.*

The performance comparison between the proposed CsiNet-GRUs to other available techniques. Where corresponding values of normalized mean square error (NMSE), ρ, accuracy, and run-time are calculated for different values of γ for indoor and outdoor scenarios, all techniques have better performances in the indoor scenario than the outdoor one.

It is worth noting that channel state information network (CsiNet) techniques significantly outperform the other CS-based methods. LASSO, TVAL3, CsiNet, and CsiNet-GRUs continue to provide the highest cosine similarity values at low CRs, where other CS-based methods fail. However, the proposed CsiNet-GRUs outperform the channel state information network (CsiNet) in terms of correlation and accuracy, as shown in **Table 2**. The same comparison is simulated again in terms of

| | | γ | **LASSO [17]** | **TVAL3 [16]** | **CsiNet [8]** | **Conv-LSTM CsiNet [18]** | **CsiNet-GRUs Dropout Epoch/1000 [14]** |
|---|---|---|---|---|---|---|---|
| Indoor | NMSE Epoch = 1000 | 1/4 | −7.59 | −8.87 | −17.36 | −27.5 | −51.73 |
| | | 1/16 | −2.96 | −3.2 | −8.65 | −23 | −27.3 |
| | | 1/32 | −1.18 | −0.46 | −6.24 | −22.3 | −23.81 |
| | | 1/64 | −0.18 | −0.6 | −5.84 | −21.2 | −22.11 |
| | ρ Epoch = 1000 | 1/4 | 0.91 | 0.87 | 0.98 | 0.95 | 0.99 |
| | | 1/16 | 0.72 | 0.73 | 0.90 | 0.93 | 0.94 |
| | | 1/32 | 0.53 | 0.45 | 0.83 | 0.85 | 0.89 |
| | | 1/64 | 0.30 | 0.24 | 0.83 | 0.84 | 0.87 |
| | Accuracy Epoch = 1000 | 1/4 | 0.68 | 0.34 | 0.81 | 0.82 | 0.84 |
| | | 1/16 | 0.55 | 0.22 | 0.60 | 0.60 | 0.62 |
| | | 1/32 | 0.34 | 0.15 | 0.51 | 0.51 | 0.52 |
| | | 1/64 | 0.55 | 0.23 | 0.48 | 0.53 | 0.54 |
| | Run time Epoch = 1000 | 1/4 | 0.345 | 0.314 | 0.0001 | 0.0001 | 0.0003 |
| | | 1/16 | 0.555 | 0.314 | 0.0001 | 0.0001 | 0.0003 |
| | | 1/32 | 0.604 | 0.286 | 0.0001 | 0.0001 | 0.0003 |
| | | 1/64 | 0.708 | 0.285 | 0.0001 | 0.0001 | 0.0003 |
| Outdoor | NMSE Epoch = 1000 | 1/4 | −5.08 | −0.9 | −8.75 | −10.9 | −15.13 |
| | | 1/16 | −1.09 | −0.53 | −4.51 | −9.86 | −11.91 |
| | | 1/32 | −0.27 | 0.42 | −2.81 | −9.18 | −3.02 |
| | | 1/64 | −0.06 | 0.74 | −1.93 | −8.83 | −2.05 |
| | ρ Epoch = 1000 | 1/4 | 0.82 | 0.58 | 0.87 | 0.90 | 0.92 |
| | | 1/16 | 0.49 | 0.46 | 0.79 | 0.81 | 0.81 |
| | | 1/32 | 0.32 | 0.28 | 0.67 | 0.68 | 0.70 |
| | | 1/64 | 0.19 | 0.19 | 0.60 | 0.62 | 0.68 |
| | Accuracy Epoch = 1000 | 1/4 | 0.66 | 0.54 | 0.68 | 0.70 | 0.71 |
| | | 1/16 | 0.45 | 0.22 | 0.49 | 0.49 | 0.51 |
| | | 1/32 | 0.20 | 0.20 | 0.36 | 0.36 | 0.38 |
| | | 1/64 | 0.18 | 0.15 | 0.26 | 0.26 | 0.27 |

| | $\gamma$ | LASSO [17] | TVAL3 [16] | CsiNet [8] | Conv-LSTM CsiNet [18] | CsiNet-GRUs Dropout Epoch/1000 [14] |
|---|---|---|---|---|---|---|
| Run time Epoch = 1000 | 1/4 | 0.2122 | 0.15 | 0.0001 | 0.0001 | 0.0003 |
| | 1/16 | 0.2409 | 0.3145 | 0.0001 | 0.0001 | 0.0003 |
| | 1/32 | 0.598 | 0.2985 | 0.0001 | 0.0001 | 0.0003 |
| | 1/64 | 0.6758 | 0.285 | 0.0001 | 0.0001 | 0.0003 |

**Table 2.**
*Comparison of results between the proposed framework and other available Ones (Epoch = 1000 iterations in the proposed techniques and others previous techniques).*

epoch = 1000 (1000 iterations) in terms of correlation and accuracy in the proposed technique CsiNet-GRUs. In terms of the NMSE, the CsiNet-GRUs achieve the lowest values of all compressed ratios (CRs), particularly when CR is low.

CsiNet-GRUs have very short run periods when compared to LASSO and TVAL3 techniques. However, when compared to the other CsiNet technique and the proposed modeling technique, CsiNet-GRUs lose time efficiency slightly. It is worth noting that, despite the addition of significant complexity as a result of the GRU layers, the run time is still comparable to that of the CsiNet.

**Figure 15** depicts in comparison to the other modeling techniques, the reconstruction results of the proposed technique, namely LASSO, TVAL3, CsiNet, and Conv-LSTM CsiNet in an indoor Picocellular scenario, the figure represents the average performance at different CRs, reflecting on the reconstructed images to use the other techniques.

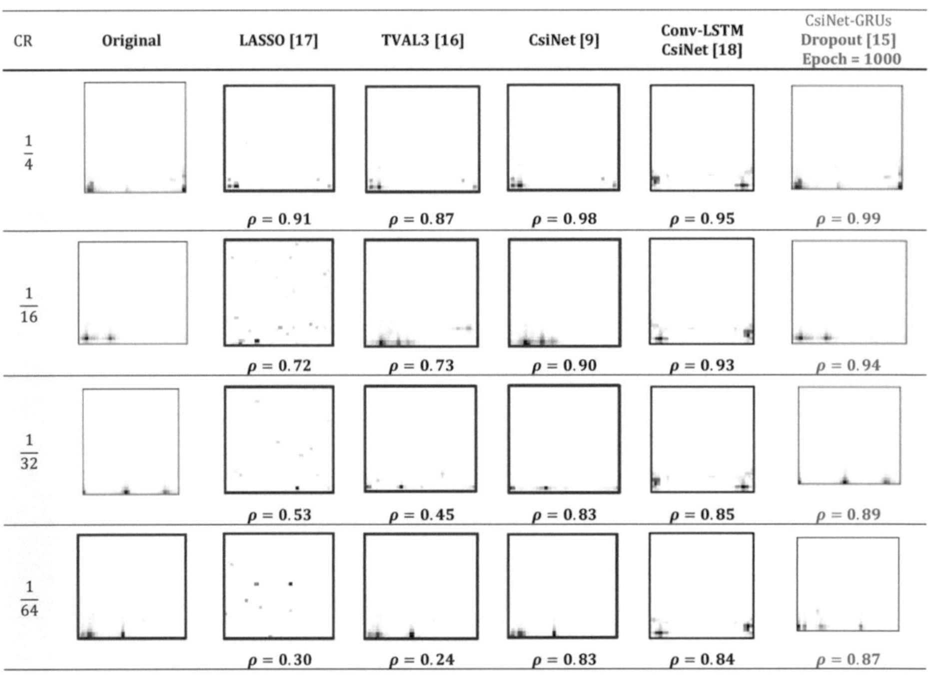


**Figure 15.**
*Reconstruction images for CR in CS algorithms in an indoor scenario.*

Conv-LSTM, CsiNet, and CsiNet-GRUs continue to provide acceptable correlation coefficients ($\rho$) at low CRs, where compressed sensing-based methods fail; it is noteworthy that the proposed CsiNet-GRUs technique outperforms the other methods in an indoor scenario. CsiNet-GRUs achieve the best performance among CR with indicators parameters to improve accuracy, decrease NMSE, and increase correlation ($\rho$) with dropout to reduce modeling system overfitting in massive multiple-input multiple-output channels. As a result, CsiNet-GRUs outperform both CsiNet and CS-based methods. With advanced deep learning technology, this chapter has the potential to deploy real MIMO systems.

## 7. Conclusion

We developed and tested a prediction model to evaluate a real-time and end-to-end channel state information (CSI) feedback framework in this chapter by extending the DL-based CsiNet with GRU. By utilizing the time correlation and structure properties of time-varying massive MIMO channels, CsiNet-GRUs achieve an acceptable trade between compressed ratio, recovery quality, accuracy, and complexity. This chapter proposed a channel state information (CSI) feedback network by extending the deep learning DL-based channel state information network (CsiNet) technique to incorporate gated recurrent units (GRUs) and use the dropout method to incorporate gated recurrent units (GRUs) and use the dropout method. The gated recurrent unit (GRU) layers were used to extend the channel state information network CsiNet decoders for time correlation extraction and the final reconstruction of channel state information, whereas the dropout method was used to reduce overfitting in channel modeling. In terms of complexity, the experiment results show that CsiNet-GRUs achieve the best recovery quality and outperform state-of-the-art CS methods.

## Appendices and nomenclature

| | |
|---|---|
| 1G | The first generation |
| 2G | The second generation |
| 3G | The third generation |
| 4G | The fourth generation |
| 5G | The fifth generation |
| AI | Artificial Intelligence |
| AMP | Approximate Message-Passing |
| AoA | Analysis of Alternatives |
| AE | Autoencoder |
| BER | Bit Error Rate |
| CE | Channel Estimation |
| CNN | Convolutional Neural Network |
| CS | Compressive Sensing |
| CSI | Channel State Information |
| CsiNet | Channel State Information Network |
| CsiNet-GRU | Channel State Information Network-Gated Recurrent Unit |
| DL | Deep Learning |
| DNN | Deep Neural Network |
| FDD | Frequency Division Duplex |

| | |
|---|---|
| GRU | Gated Recurrent Unit |
| LASSO | Least Absolute Shrinkage and Selection Operator |
| LSTM | Long Short-Term Memory |
| MIMO | Multiple-Input Multiple-Output |
| mmWave | Millimeter Wave |
| MSE | Mean Squared Error |
| NMSE | Normalized Mean Square Error |
| RELU | Rectified Linear Unit |
| RNN | Recurrent Neural Network |
| SNR | Signal-to-Noise-Ratio |
| $(\cdot)^H$ | Conjugate transpose |
| $\rho_g$ | The original is a sigmoid function |
| $\phi_h$ | The original is a hyperbolic tangent |
| $\gamma$ | Compression Ratio |
| $\theta$ | The delay |
| $\partial t$ | Feedback internal |
| $\|\cdot\|_2$ | The Euclidean norm |
| $\mathbb{E}$ | Exponent |
| $\mathbb{C}$ | Complex numbers |
| $\sum$ | Summation |
| € | Element |
| $\tilde{h}_n^H$ | The channel frequency response vector at the $n^{th}$ subcarriers |
| $h_t$ | Output vector |
| $L(\theta)$ | Loss Function |
| $N_c$ | Receiver antenna at the user equipment |
| $N_t$ | Transmit antenna at the base station |
| $r_t$ | Reset gate vector |
| T | Time sequence of the channel model |
| $x_n$ | The transmitted information image |
| $x_t$ | Input vector |
| $y_n$ | The pre-coding vector at the $n^{th}$ subcarriers |
| $z_n$ | The additive noise or obstruction |
| $z_t$ | Update gate vector |

**Author details**

Hany Helmy[1*], Sherif El Diasty[2] and Hazem Shatila[3]

1 Cairo Airport Company (CAC), Cairo, Egypt

2 Department of Electronics, Arab Academy for Science, Technology and Maritime Transport (AASTMT), Cairo, Egypt

3 Virginia Tech, Artificial Intelligence and Markovdata, Cairo, Egypt

*Address all correspondence to: hany.nabil@cairo-airport.com; hnabil110@gmail.com

## References

[1] Zhang T, Ge A, Beaulieu NC, Hu Z, Loo J. A limited feedback scheme for massive MIMO systems based on principal component analysis. EURASIP Journal on Advances in Signal Processing. 2016;**2016**. DOI: 10.1186/s13634-016-0364-9

[2] Busari A, Huq KMS, Mumtaz S, Dai L, Rodriguez J. Millimeter-wave massive MIMO communication for future wireless systems: A survey. IEEE Communications Surveys & Tutorials. 2018;**20**(2):836-869

[3] Tao J, Chen J, Xing J, Fu S, Xie J. Autoencoder neural network based intelligent hybrid beamforming design for mmWave massive MIMO systems. IEEE Transactions on Cognitive Communications and Networking. 2020. DOI: 10.1109/TCCN.2020.2991878

[4] Zhai J, Zhang S, Chen J, He Q. Autoencoder and Its Various Variants. In: 2018 IEEE International Conference on Systems, Man, and Cybernetics (SMC), Miyazaki, Japan. 2018. pp. 415-419. DOI: 10.1109/SMC.2018.00080

[5] Karanov B, Lavery D, Bayvel P, Schmalen L. End-to-end optimized transmission over dispersive intensity-modulated channels using bidirectional recurrent neural networks. Optics Express. 2019;**27**:19650-19663

[6] Sohrabi F, Cheng HV, Yu W. Robust Symbol-Level Precoding Via Autoencoder-Based Deep Learning. 2020. pp. 8951-8955. DOI: 10.1109/ICASSP40776.2020.9054488

[7] Liu Z, del Rosario M, Liang X, Zhang L, Ding Z. Spherical Normalization for Learned Compressive Feedback in Massive MIMO CSI Acquisition. 2020. pp. 1-6. DOI: 10.1109/ICCWorkshops49005.2020.9145171

[8] Wen C, Shih W, Jin S. Deep learning for massive MIMO CSI feedback. IEEE Wireless Communications Letters. 2018; **7**(5):748-751

[9] Liu L, Oestges C, Poutanen J, Haneda K. The COST 2100 MIMO channel model. IEEE Wireless Communications. 2012;**19**(6):92-99

[10] Hochreiter S. The vanishing gradient problem during learning recurrent neural nets and problem solutions. International Journal of Uncertainty, Fuzziness and Knowledge-Based Systems. 1998;**6**(2):107-116

[11] Aleem S, Huda N, Amin R, Khalid S, Alshamrani SS, Alshehri A. Machine Learning Algorithms for Depression: Diagnosis, Insights, and Research Directions. Electronics. 2022;**11**(7):1111. DOI: 10.3390/electronics11071111

[12] Cho K, van Merrienboer B, Gulcehre C, Bahdanau D, Bougares F, Schwenk H, et al. Learning Phrase Representations using RNN Encoder-Decoder for Statistical Machine Translation. 2014. DOI: 10.3115/v1/D14-1179

[13] Dey R, Salem FM. Gate-variants of Gated Recurrent Unit (GRU) neural networks. 2017. pp. 1597-1600. DOI: 10.1109/MWSCAS.2017.8053243

[14] Helmy HMN, Daysti SE, Shatila H, Aboul-Dahab M. Performance enhancement of massive MIMO using deep learning-based channel estimation. IOP Conference Series: Materials Science and Engineering. 2021;**1051**(1):012029

[15] Srivastava N, Hinton G, Krizhevsky A, Sutskever I, Salakhutdinov R. Dropout: A simple way to prevent neural networks from overfitting. Journal of

Machine Learning Research. 2014;**15**(1):1929-1958

[16] Li C, Yin W, Zhang Y. User's guide for TVAL3: TV minimization by augmented Lagrangian and alternating direction algorithms. CAAM Report. 2009;**20**(4):46-47

[17] Daubechies I, Defrise M, Mol CD. An iterative thresholding algorithm for linear inverse problems with a sparsity constraint. Communications on Pure and Applied Mathematics. 2004;**75**:1412-1457

[18] Li X, Huaming W. Spatio-temporal representation with a deep neural recurrent network in MIMO CSI feedback. IEEE Wireless Communications Letters. 2020;**9**(5):653-657

Chapter 2

# Graph Neural Networks and Reinforcement Learning: A Survey

*Fatemeh Fathinezhad, Peyman Adibi, Bijan Shoushtarian and Jocelyn Chanussot*

## Abstract

Graph neural network (GNN) is an emerging field of research that tries to generalize deep learning architectures to work with non-Euclidean data. Nowadays, combining deep reinforcement learning (DRL) with GNN for graph-structured problems, especially in multi-agent environments, is a powerful technique in modern deep learning. From the computational point of view, multi-agent environments are inherently complex, because future rewards depend on the joint actions of multiple agents. This chapter tries to examine different types of applying GNN and DRL techniques in the most common representations of multi-agent problems and their challenges. In general, the fusion of GNN and DRL can be addressed from two different points of view. First, GNN is used to influence the DRL performance and improve its formulation. Here, GNN is applied in relational DRL structures such as multi-agent and multi-task DRL. Second, DRL is used to improve the application of GNN. From this viewpoint, DRL can be used for a variety of purposes including neural architecture search and improving the explanatory power of GNN predictions.

**Keywords:** graph neural network, deep reinforcement learning, multi-agent, multi-task, neural architecture search

## 1. Introduction

Building an intelligent system that can extract high-level representation from data is necessary for many issues related to artificial intelligence. Theoretical and biological arguments show that to build such systems, deep architecture models are needed that include many layers of non-linear processing units. Before the emergence of deep learning [1], traditional machine learning approaches depended on the representations given by feature selection or extraction that get from the data.

These methods required an expert in the domain of the subject to extract the features manually. However, this hand-crafted feature extraction is a time-consuming and sub-optimal process. The emergence of deep learning could quickly replace these traditional methods because it could automatically extract the features according to each problem. In recent years, deep learning has become the main motivation for innovative solutions to artificial intelligence problems. This issue has been made possible by increasing the amount of available data, increasing computing resources, and improving techniques in training deep networks.

Deep neural networks (DNNs) have reached remarkable achievements in the last decade [2]. However, basic types of neural networks can only be implemented using regular or Euclidean data. Whereas, many data in the real world have a graph structure that is a non-Euclidean data structure. This irregularity of the data structure has led to recent advances in graph neural networks (GNNs).

GNNs [3, 4] allow the creation of a machine learning model which is taught simultaneously to learn the representation of data with a graph structure. GNNs are undoubtedly the most interesting issue in graph-based deep learning. GNNs can be applied to graph-structured data for various tasks, from clustering to classification or regression. They can also learn representation at the level of nodes, edges, and graphs. Deep learning with graph data structure is recognized as graph representation learning, geometric deep learning, or graph embedding which seeks to learn the representation of structured information about graphs.

The purpose of graph representation is to build sets of features that show the structure of the graph and the data in it. The main key of this method is to learn a mapping that embeds nodes or graphs as points in a low-dimensional vector space so that this mapping is optimized and the geometric relations in this learned space reflect the structure of the original graph. After optimizing the built-in space, this learned embedding space can be used as input features of the graph.

Once the size of the dataset in the input network is different from the training history GNNs play a highly efficient role in knowledge transfer between data-oriented structures. GNNs are inherently designed to generalize over graphs of different structures and sizes. This ability allows the GNN-based DRL agent to learn and generalize over arbitrary network of environment topologies. Many DRL methods apply standard neural networks such as recurrent neural networks (RNNs) [5], and other neural network structures. This issue causes poor generalization and prevents the deployment of DRL in networks, because it is hard to adjust to the dynamic changes of network topology. In recent years, the integration of GNNs and DRL specially in multi-agent systems has attracted much attention in graph-structured environments.

Nowadays, many systems can be viewed as multi-agent systems from a new perspective. The cooperation of a group of agents (teamwork) in the frame of a graph is one of the most important issues that is always raised in multi-agent systems [6], due to increasing the ability to reach the final goal of the system and improving the overall strategy.

This issue becomes more important when the environment is complex and dynamic. In such an environment, a purposeful agent is affected by the actions of other agents in addition to the changes in the environment. Therefore, the environment has more dynamic states than before, and the agent must have the ability to model the action process and the power to learn and interact with other agents. Using classical methods in describing agents and establishing communication between them in a multi-agent environment, due to the use of many equations, weakens the power of expanding the network to large systems. By defining an intelligent system and using smart modern methods in solving such problems, methods such as deep reinforcement learning algorithms have been proven to be useful in such environments.

Automated control problems and development in decision-making are the results of recent advances in DRL [7]. However, existing DRL-based solutions still fail to generalize when applied to network-related scenarios. So, when faced with a network state that is not seen during training, the ability of the DRL agent is impaired.

Recently, GNNs have been offered to model and operate on graphs to reach combinatorial generalization and relational reasoning. Indeed, GNNs simplify the learning of relations between entities in a graph and the rules for composing them. A combination of DRL and GNN can work and optimize problems while generalizing to unseen topologies. Specifically, the GNN used by the DRL agent is inspired by message-passing neural networks [8].

Robotics, pattern recognition, recommendation systems, and games are some of the subjects in which DRL has presented acceptable performance. On the other hand, GNNs exhibit excellent efficiency in supervised learning for graph-structured data [9]. DRLs utilize the ability of DNNs to solve sequential decision problems with RL, and on the other hand, GNNs are new architectures that are suitable for organizing graph-structured data in this field.

In this survey, an overview of the concepts of GNNs is prepared, and then their relationship with reinforcement learning (RL) is explained. The rest of this chapter is structured as follows. A short review of graph neural networks is given in Section 2. The technical backgrounds of deep reinforcement learning concepts and multi-agent reinforcement learning are presented in Section 3. The relation between RL and GNN is presented in Section 4. Finally, the conclusion is provided in the last section.

## 2. Graph neural networks

Nowadays, many learning problems need to use graph representation to present the complex relationship between data [10, 11]. Recently, more attention over studies on graph models has been received due to the great expressive power in social science (social networks) [12–14] and biology science (predicting protein interface and bioinformatics analysis, knowledge graphs, modeling physics systems, and classifying diseases) [15–17].

Pairwise message passing is one of the main elements in the structure of GNNs, such that each node in the graph frequently updates its representations by replacing information with its neighbors until a stable balance is attended. The graph neural network usually contains two parts: the message passing part for extraction of local infrastructure features used around the nodes and the readout phase which is an aggregation part to summarize the particular features of the node in a vector of features of the graph surface.

Representing data as a graph has several advantages, such as a simplified representation of complex problems, systematic modeling of relationships, etc. On the other hand, working with data with a graph structure using common DNN-based methods has its own challenges. The variable size of the unordered nodes, the uneven structure of the graph, and the dynamic neighborhood composition make it difficult to implement basic mathematical methods such as convolution on the graph. Graph neural networks (GNN) as its general structure is shown in **Figure 1**, overcome this defect with the help of new DNN methods in the graph structures of datasets. GNN architectures can model structural information and node features. In the following several well-known models of graph neural networks are introduced.

### 2.1 Graph convolutional network

For the first time in [18], spectral networks and local deep networks were connected on a graph convolutional network (GCN), as a method for semi-supervised

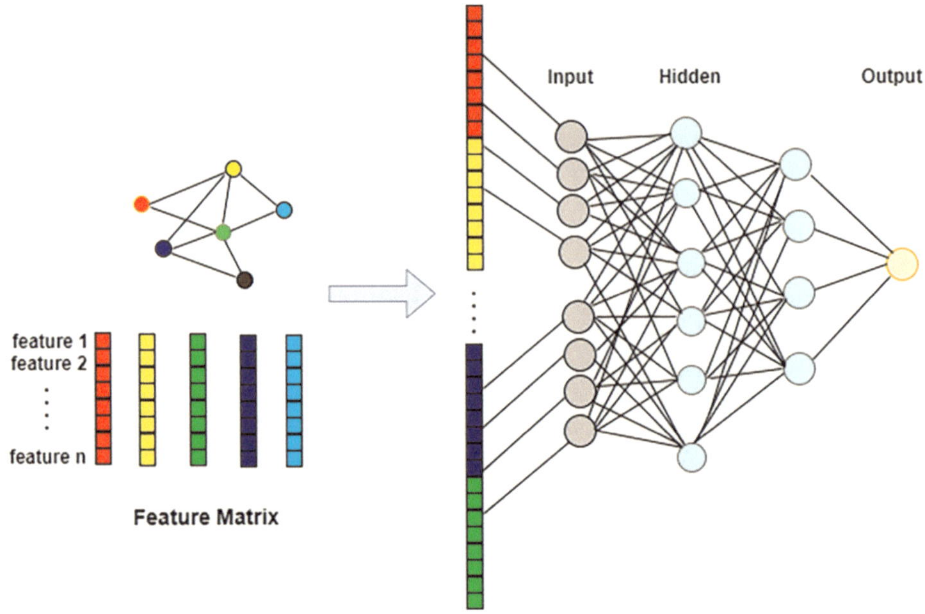


**Figure 1.**
*Graph neural networks (GNN) framework.*

learning on graph-structured data. The definition of these networks is based on the notion of convolutional neural networks, which are applied to the graphs. GCNs [19] are learned according to the structure of the features of the neighboring nodes. In general, the main difference between CNN and GCN is related to their data structure. CNNs are defined in Euclidean space while GCNs work on graph structure (non-Euclidean structure data) where the number of node connections is different and also the nodes are unordered.

Spatial graph convolutional networks and spectral graph convolutional networks are the two main branches of GCNs. The key idea in spectral GCN was defined by signal/wave propagation. In spectral GCN, information is propagated along the nodes as signal propagation. Eigen-decomposition of graph Laplacian matrix in spectral GCNs is used for information propagation and also is used for node classification by understanding the graph structure. Non-generalization and inefficiency of computations in spectral graph convolutional networks are two main challenges in spectral graph convolutional networks. GCN overcomes these problems by Chebyshev polynomials to approximate the spectral convolution and the ChebNet network is defined [20].

## 2.2 Graph attention network

Graph attention network (GAT) architecture [21] is a type of GCN architecture in which the aggregation process learns the weights between the neighboring nodes of each node with the help of the attention mechanism. In these networks, greater weights are applied to more important nodes and it stores the weight of the nodes. The advantage of graph attention networks is that these networks can adaptively learn the importance of each neighbor. However, since the attention weights between each pair of neighbors must be calculated, the calculation cost as well as the amount of memory occupied increases rapidly.

### 2.3 GraphSAGE

In graph theory, there is a concept called node embedding, which means mapping nodes to an embedded space with dimensions less than the actual dimension of the data defined on the nodes of the graph, in which similar nodes are embedded close to each other, in the resulting latent space.

GraphSAGE [22] is a deductive learning technique that exploits node features to learn an embedding function for dynamic graphs. This inductive learning approach is scalable across graphs of different sizes as well as subgraphs within a given graph. A new node can be embedded without retraining by the GraphSAGE approach. It uses aggregator functions to induce new node embeddings based on node features and neighborhoods.

In [23] a method for data-driven neighborhood subsampling is defined by a non-linear regressor based on the real-valued importance of each node and its neighborhood. This subsampling helps to embed nodes in the graph using a small set of neighboring nodes with high importance. The regressor is learned using value-based reinforcement learning. Here, the negative classification loss output of GraphSAGE is used to extract this importance.

GraphSAGE-D3QN [24] presents a graph DRL method for emergency control of undervoltage load shedding model. Feature extraction of states in this model is designed by GraphSAGE-based method with topology variation in the training step and then online emergency control is achieved.

### 2.4 Applications of GNNs

Link prediction [25], node classification [26], clustering [27] and, etc., are considered as graph analysis objectives. In the following, several common GNNs goals are described:

**Node classification**: training models to classify nodes by determining the label of samples that are shown as nodes. Usually, these problems are used in a semi-supervised way, with only a part of the graph being labeled.

**Graph classification**: Graph classification is a task with real applications in social network analysis, categorizing documents in natural language processing, and classifying proteins in bioinformatics fields. Graph classification obtains a graph feature that aids discriminate between graphs of different classes.

**Graph Visualization:** Visual representation of data structures and anomalies with the help of geometric graph theory and information visualization that helps the user understand graphs.

**Link prediction:** Predicting the relationship between two nodes and considering that nodes in a network are likely to have links. An application of this approach is to detect social interactions or suggest potential friends to users on social networks. It has also been used in predicting criminal associations, and in recommender system problems.

**Graph clustering:** clustering on graphs is performed in two ways. Either clustering is based on nodes that should be converted into different and connected groups based on the edge distances and their weights or considering the graph as objects that should be clustered, and clusters these objects based on similarity.

## 3. Deep reinforcement learning

Using DNNs to solve sequential decision issues in the framework of RL led to the emergence of deep reinforcement learning (DRL) in high-dimensional problems (see **Figure 2**). Nowadays, different applications of artificial intelligence have been enhanced with the help of DRL which includes natural language processing [28], transportation [29], finance [30], healthcare [31], robotics [32], recommendation systems [33], and gaming [34]. DRL can be defined as a system that maximizes the long-term reward in a reinforcement learning problem using representations that are themselves learned by the deep network. The outstanding success of DRL can be considered due to the ability of this method to deal with complex problems and provide efficient, scalable, and flexible computational methods. Also, DRLs have a high ability to understand the dynamics of the environment and produce optimal actions according to their interactions with the environment. When dealing with various high-dimensional problems or continuous states, reinforcement learning suffers from the problem of inefficient feature representation. Therefore, learning time is slow and techniques should be designed to speed up the learning process. The most significant feature of deep learning is that DNNs can discover compact representations of high-dimensional data automatically.

Combining DNNs with RL has become more attractive in recent years and it has gently shifted the focus from single-agent environments to multi-agent ones. Working

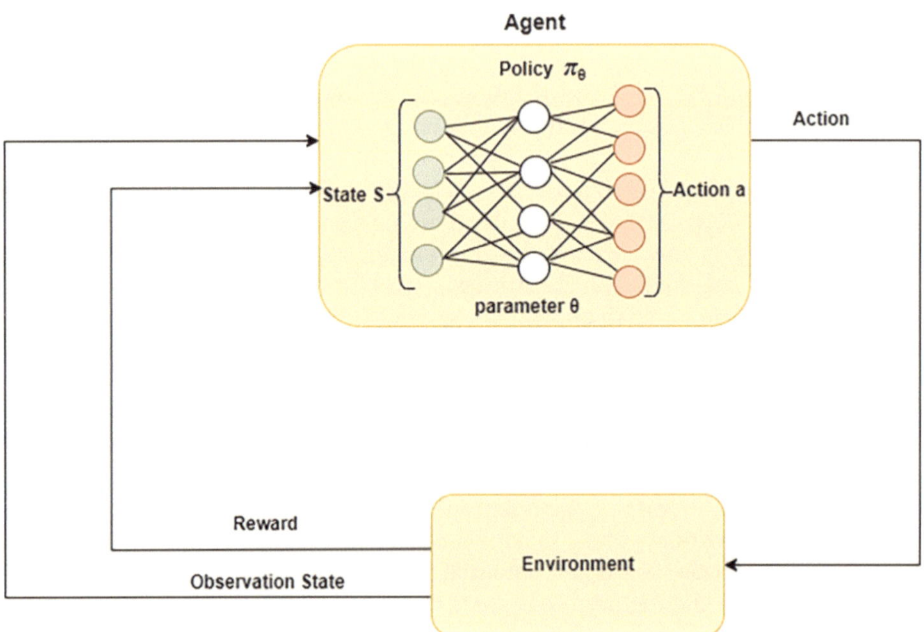


**Figure 2.**
*Total structure of the combination of GNNs and DRL.*

with multiple agents is inherently more complex because future rewards depend on the joint actions of several agents and the computational complexity of the function increases. Single-agent environments such as Atari [6], and navigation robots [35], and multi-agent settings such as traffic light control [36], financial market trading [37], and strategy games such as Go, StarCraft, and Dota are some examples that are developed by DRL.

In DRL, unstructured input data from the state space are given to the network. This input such as pixels rendered on the screen in a video game or images from a camera or the raw sensor stream from a robot can be very large and high-dimensional. In the output, the value of an action is determined for the agent to decide what actions must be performed in the environment to maximize the expected rewards. Since the RL methods are suffered from the curse of dimensions problem. DNNs can find low-dimensional representations (and features) of high-dimensional data automatically. In the following, the subject of DRL for the special scope of multi-agent reinforcement learning will be expressed widely.

### 3.1 Multi-agent reinforcement learning

In multi-agent reinforcement learning (MARL) sets of independent agents interact with each other to learn how to reach their goals. Large and random state spaces are a common problem in MARL systems with dynamic environments. These challenges include inefficient cooperation between agents, unsuitable coordination between agent decisions, and the effect of state space size on the learning time. In recent years, MARL applications have been used in autonomous driving, traffic light control, and network packet delivery. Communication between agents gathers information about the environment and the status of other agents.

Markov decision process (MDP) is a useful approach for modeling optimal decision-making in stochastic environments such as multi-agent environments but with different representations. The dynamics of the state and the expected rewards change for agents according to the common action and violate the stationarity assumption of MDP. MDP can be completely or partially observable in a multi-agent environment. It also depends on the type of interaction, which can be competitive, collaborative, or mixed. They perform actions sequentially or simultaneously.

Markov game [38], represent a theoretical framework for the study of agents with multiple interactions in a fully observable environment and can be used in competitive, cooperative, and hybrid environments. A Markov game is a set of regular games (matrix games) that agents perform repeatedly in it. Each state of the game can be represented as a matrix representation with the payoff of each joint action. If the agents cooperate with each other; but the actions have decentralized execution, it is shown by a decentralized MDP.

A partially observable Markov game is a multi-agent Markov decision process in which every agent has an individual partial observation of the environment and takes an individual action to receive their own reward. If the agents cooperate to optimize a single reward function according to the joint state and action, then the problem can be modeled as a decentralized partially observable Markov decision process (Dec-POMDP). RL in a multi-agent space is associated with several problems. Partial observability, non-stationarity, computational complexity, and credit allocation are among these problems. In the following, each of these aspects will be discussed:

### 3.1.1 Partially observable

Based on local observations in a partially observable environment, each agent makes decisions. So, it leads to asymmetric and incomplete information among agents, which makes the learning process difficult. Partially observable working has been studied mainly in situations where a group of agents maximizes team rewards through a common policy. For example, in Dec-POMDP settings, the two main approaches are (1) centralized learning and decentralized execution paradigm, and (2) using communication to exchange information about the environment.

Since in Dec-POMDP the agents partially observe the state and try to maximize the rewards in each step for all agents, the optimal solution for a Dec-POMDP model is considered a challenge. The lack of access to the real state information in the Dec-POMDP leads to the use of the history of observations and actions, which is computationally expensive for solving the Dec-POMDP model. Policy tree by pruning suboptimal trees [39, 40], and a feature-based heuristic search value iteration [41] techniques are used to solve this challenge in Dec-POMDP model.

Also, deep multi-agent reinforcement learning algorithms for Dec-POMDP models are considered an approximate policy solution technique. Different MARL algorithms have been represented to produce decent policies on many challenging dec-POMDP problems [42, 43]. An independent learning approach is used in [43] where a policy solution for each agent is updated solely based on their individual experiences.

### 3.1.2 Non-stationary

In a multi-agent environment, all agents simultaneously learn, interact and change the environment. As a result, state transitions and rewards are no longer fixed, and agents continue to adapt to the changing policies of other agents. This violates the Markov assumption, which is problematic because most RL algorithms assume a fixed environment to guarantee convergence. One way to deal with non-stationarity is to learn as much as possible about the environment, e.g., through adversary modeling and information exchange between agents [44].

To solve the non-stationarity problem the centralized critic architecture is used. Actor-critic algorithm for this architecture includes two components. The critics' training is centralized and has access to the observations and actions of all agents, while the actors' training is decentralized. Since the actor computes the policy, the critic component can be removed during testing, and therefore the approach has fully decentralized execution. If the actions and observations of the opponent during the training are available, the agents do not experience unexpected changes in the dynamics of the environment and it will lead to the stabilization of the process.

The actor-critic algorithm in [42] is used with stochastic policies to evaluate and train agents in the StarCraft game. A single centralized critic is applied for all the agents and a different actor for each agent is used.

Generally, considering non-stationarity in multi-agent systems does not need centralized training approaches. Self-play is another decentralized method that has been explored to manage non-stationarity in MARL problems. This approach trains a neural network, using each agents' observation as input, by playing it against its current or previous versions to learn policies that can generalize to any opponent.

#### *3.1.3 Computational complexity*

As each agent is added, state-action space grows exponentially, which leads to an increase in the time complexity of algorithms in multi-agent environments. Training a DRL model for a single-agent needs significant sources and gets worse for several interaction agents which leads to slow learning.

Reducing the learning complexity for goal-directed problems can be achieved by initializing the Q-values with a good approximative function. In multi-agent problems, good approximations for a big class of problems, namely for goal-directed stochastic games, exists [45]. These games can reflect coordination cooperative robotics.

#### *3.1.4 Assignment of credit*

Allocation of credit to agents due to the simultaneous performance of several agents in the environment leads to the difficulty of learning an optimal policy in the environment. The individual contribution of an agent cannot be determined in the joint reward signal [46]. The agent is also able to distinguish whether changes in global reward are due to its actions or those of other agents. One way to solve this problem is to let agents learn based on a local reward. But the agent may easily increase his local reward, which encourages selfish behavior that may reduce overall group performance. Several approaches are discussed which were created to deal with these challenges.

### 3.2 Interaction methods between multi-agents with GNN architecture

In the most recent research, many MARL methods use GNNs to provide information interactions between agents to complete collaborative tasks and coordinate actions. In general, not extracting enough useful information from neighboring agents is one of the problems of simply aggregation in GNN, which is due to ignoring the topological relationships in the graph.

To solve this problem, Ding et al. [47] presented a method to extract useful information from neighboring agents as much as possible in the graph structure, which has the ability to provide feature representation to complete the cooperation task. For this purpose, mutual information (MI) is applied for measuring the agent topological relationships and the agent features information to maximize the correlation between input feature information of neighbor agents and output high-level hidden feature representations.

A GNN architecture for training decentralized agent policies on the perimeter of a unit circle has been proposed in continuous action spaces [48]. In this approach, multi-agent perimeter defense problems are solved by learning decentralized strategies with GNNs. Local perceptions of the defenders are considered as inputs in the learning framework and finally, the model is trained by an expert policy based on the maximum matching algorithm and returns actions to maximize the number of captures for the defender team.

The proposed framework [49] used GNNs for value function factorization in multi-agent deep reinforcement learning. A complete directed graph is designed by the team of agents as the nodes of the graph, and edge weights are determined by an attention mechanism. The introduced mixing GNN (GraphMIX) module in this paper is responsible for factorizing the team value function into individual per-agent observation-action value functions, and explicit credit assignment to each agent. The centralized-training-decentralized-execution in GraphMIX give the ability to the agents to make their decisions independently once training is completed.

An attention mechanism in [50] is defined to adjust the weights of the edges during an episode based on the agents' observation-action histories. To create the factorized state-action value function's implicit assignment of global reward, additional per-agent loss terms are taken from the output node embeddings of the GNN, that divide the global reward to individual agents explicitly. Neural attention modules have been used in the graph structures [50, 51], for applying attention mechanisms to compute graph edge weights. These techniques are used in sentence translation works for managing associations between structured data [52], and they are generally used in RL [53].

Non-stationery and coordination problems can be solved naturally by centralized learning of joint actions but it is difficult to scale, because of the exponentially grows of the joint action space by the number of agents. To solve this challenge, conditional independencies between agents are exploited by decomposing a global reward function into a sum of agent-local terms. Sparse cooperative Q-learning [54] is a tabular Q-learning algorithm that learns to coordinate the actions of a group of cooperative agents only in the states in which such coordination is necessary, encoding those dependencies in a coordination graph. The use of these methods requires the prior provision of dependencies between agents. To solve this problem, it is assumed that each agent always contributes to the global reward and learns the amount of its contribution in each state.

Coordinating graph formulation is one of the methods for determining the joint action between agents based on the structure of interactions. In [55], Deep Implicit Coordination Graph (DICG) architecture is introduced, which includes two modules, one for obtaining the dynamic coordination graph structure and the other for learning the implicit reasoning about common actions or values. DICG uses the actor-critic structure to improve coordination for multi-agent situations. DICG is assumed that agents can pass messages that encode their observations. The agents use GCN to pass these messages between one another, where the adjacency matrix for the network is learned with self-attention. Here, a new state categorization method has been presented for centralized-training-decentralized-execution. In this method, which is implemented in the StarCraft game, each game agent separates information and observations of itself and its competitors and then leverages GAT to learn the correlation and relationship among the agents.

### 3.3 Different methods for computing value function in MARL

This section describes different methods for calculating the Q-value function for multi-agent environments. In MARL problems, each agent has a local and private observation of its surrounding space that it wants to take action based on that information. A problem that the agent may face with it is the locality of observation and not having complete information about the environment. Another problem is the non-stationarity of the environment because all agents in the environment are learning and show different behaviors during training.

To solve these problems, the simplest method is to use single-agent RL algorithms for each agent and consider other agents as part of the environment. However, the exponential growth of this joint action space becomes difficult with the number of agents. The Independent Q-Learning (IQL) [56] method is based on this logic and has a good efficiency in some multi-agent RL problems, but there is no guarantee of their convergence. In IQL, each agent has a separate action value function based on which it receives the local observation of the agent and then chooses its action based on it. In such environments, RNN can also be used for the history of observation-action.

In another approach, the agents perform learning in a centralized manner and the choice of action is also centralized. This approach is suitable for problems (such as traffic management or traffic light management) that do not require decentralized execution.

The third approach includes centralized training and decentralized execution. In this approach, the agents have access to the state and complete information during the training step, but in some environments, the learned policy must be applied in a decentralized manner, and the agents cannot access the full state in the execution phase. In this method, the purpose of each agent is to perform actions that maximize their utility function (joint value function), but such decentralization can result in sub-optimal actions [55].

Value-based methods like Value Decomposition Networks (VDN) [57], QMIX [58] and actor-critic methods like Multi-Agent Deep Deterministic Policy Gradient (MADDPG) [59] and Counterfactual Multi-Agent (COMA) [60] are some approaches presented to solve these problems with training in a centralized manner and execution in a decentralized way.

In VDN, a linear summation of all action-value functions of all agents is used to determine separate action-value functions for each agent and learn using only a common reward signal. Using a common reward signal, it tries to learn the decomposed value functions for each agent and use it for decentralized execution. QMIX generalized the VDN method and combines the Q-value of different agents in a non-linear way. They use the global state as input to hyper networks to generate weights and biases of the mixing network. The actor-critic architecture is the basis of centralized training and decentralized execution. In this method, they use the full state and additional information available in the training phase of the critical network to generate a richer signal for the actor.

One of the disadvantages of the aforementioned above algorithms is that it does not clearly obtain the underlying structure of cooperation between agents with a graph topology. Some papers try to join MARL with graph learning. For example, a multi-agent deep reinforcement learning based on GCN structure has been presented [61]. Here, the decentralized decision-making is not considered by the agents and only centralized training and centralized execution are investigated for communicating agents with each other during the inference phase several times.

Multi-agent DDPG (MADDPG), generalizes the actor-critic algorithm into a multi-agent policy gradient algorithm where decentralized agents learn a centralized critic based on the observations and actions of all agents. It leads to learned policies that only use local information and observations at execution time. This method does not assume a differentiable model of the environment dynamics or any particular structure of the communication method between agents. It applies not only to cooperative interaction but also to competitive or mixed interaction involving both physical and communicative behavior. The critic is strengthened by additional information about other agents' policies, while local information is provided just for the actor. After training completion, only the local actors are used in the execution phase, acting in a decentralized manner.

COMA is a multi-agent policy gradient-based method for cooperative multi-agent systems that uses a centralized critic to estimate Q performance and decentralized actors to optimize agent policies. Also, this method solves the problem of credit assignment using a count. Unlike COMA, which uses a centralized critic for all agents, MADDPG has a concentrated critic for each agent to have different reward functions in competitive environments.

Recent works have been conducted based on MADDPG, R-MADDPG [62] develops the MADDPG algorithm to the semi-observable environment by preserving the history of previous observations in the critic module and by having an iterative actor. M3DDPG [63] includes minimax optimization for powerful policy learning against agents with changing strategies. Actor-Critic with mean field [64] factorizes the Q-value function only by using interaction with neighboring agents based on mean field theory, and the idea of dropping out can be expanded to MADDPG for managing large input space [65].

## 4. Combination of graph neural networks and reinforcement learning

Recently, combining GNNs with reinforcement learning for graph-structured problems is a powerful tool in modern deep learning [66]. Combinatorial optimization [67], transportation problems [68], and manufacturing and control [69] are interesting applications in these fields.

**Figure 3** shows the total structure of the combination of GNNs and DRL. The local observation of agents is encoded by MLP for low-dimensional input or CNN for visual input into the feature vector which is shown in the embedded layer. The attention

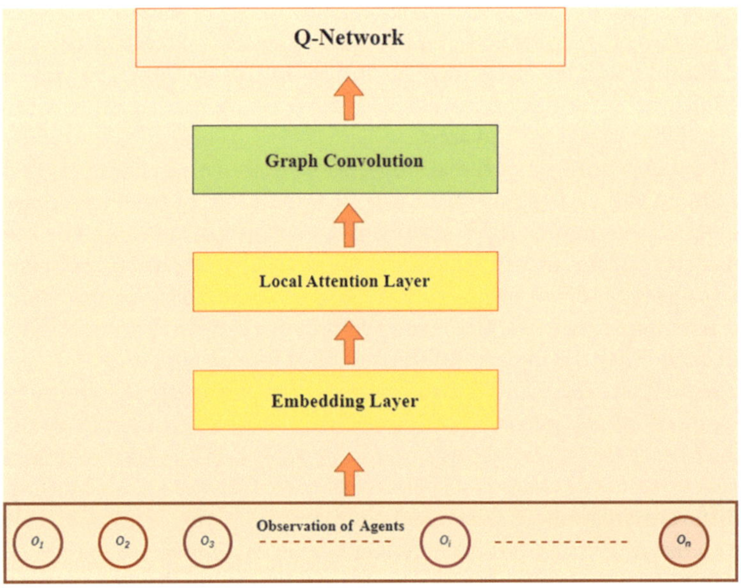


**Figure 3.**
*Schematic structure of deep reinforcement learning agent.*

network usually represents to define the edge weights as the strength of the connection in the coordination graph between each agent and its neighbors. In the next step, the graph convolution layer is applied to perform message passing and information integration across all agents. Finally, the deep Q-network is used to approximate the Q-value function. By considering the maximum output of the Q-network the next action for the agents is determined.

The embedding layer contains an encoder for $n$ observations $\{o_1, o_2, \dots, o_n\}$ of $n$ agents. The outputs of the encoder include embedding vectors $E_i$ for $i = 1..n$ as follows:

$$E_i = \text{Encoder}(o_i, \theta_E) \tag{1}$$

In the local attention Layer, the attention weights for two agents $i$ and $j$ in the graph are calculated using embedding vectors as:

$$At_{ij} = \frac{\exp(\text{Attention }(\mathrm{E}_i, \mathrm{E}_j, \mathbf{W}_a))}{\sum_{k=1}^{n} \exp(\text{Attention }(\mathrm{E}_i, \mathrm{E}_k, \mathbf{W}_a)} \tag{2}$$

where the attention network is parametrized by the weight matrix $\mathbf{W}_a$.

Message passing and information integration across all agents are expressed in a graph convolution layer as follows:

$$\mathrm{H}^{(l+1)} = \sigma\left(\tilde{\mathbf{D}}^{-\frac{1}{2}} \tilde{\mathbf{A}} \tilde{\mathbf{D}}^{-\frac{1}{2}} \mathbf{H}^{(l)} \mathbf{W}_c{}^{(l)}\right) \tag{3}$$

where $\mathrm{H}^{(l)}$ is the feature matrix of convolution layer $l$, $\tilde{\mathbf{A}} = \mathbf{A} + \mathbf{I}_N$, and $\tilde{D}_{ii} = \sum_j \tilde{A}_{ij}$.

The predicted $\hat{Q}$ in Q-network is verified by $\theta$ parameter. The general objective for each minibatch in the training step is to minimize the loss function as:

$$L_\theta = \frac{1}{b} \sum_t y_t - \hat{Q}\left(s_t, a_t, \theta_{predict}\right) \tag{4}$$

where $b$ is the batch size, and $y_t = r_{t+1} + \gamma \max_{a_{t+1}} Q\left(s_{t+1}, a_{t+1}, \theta_{target}\right)$ in time step $t$ is the target of Q value function for state $s$ and action $a$ with reward $r$.

In general, the combination of GNN and DRL can be addressed from two different points of view. From one perspective, GNN is used to advance the formulation and performance of DRL and specifically, when GNN has been used for relational DRL problems. The successful modeling for this relationship can be defined among (1) different agents in a multi-agent deep reinforcement learning (MADRL) framework, and (2) different tasks in multi-task deep reinforcement learning (MTDRL) framework [70].

From another perspective, DRL can be used to progress the performance of GNN. DRL is used to improve the explanatory power of GNN predictions, Neural Architecture Search (NAS) [71], and design adversarial examples for GNN. NAS is the process of automatically searching for the optimal architecture of a particular neural network to solve a problem, which includes finding the number of layers, the number of nodes in the layer, etc. In GraphNAS [72], the RL algorithm helps to search in the graph neural architectures. GraphNAS represents a search space for covering sampling functions, aggregation functions, and gated functions. To define the architecture of a graph neural network a recurrent network is used to create variable-length strings. Auto-GNN [73] is defined in the predefined search space by RL-based controllers. This architecture is applied in the hidden dimension, attention head, attention function, activation function, and aggregate function.

Identifying the subgraph that can have the most impact on the prediction process in GNN is one of the problems in generating explanations for GNN predictions, and in [74], DRLs are used for this improvement. Here, a DRL-based iterative graph generator is used the most important node for a prediction as a seed node is selected and then adds edges to generate the explanatory sub-graphs.

Learning a sub-graph generation policy with a policy gradient is done by mutual information of predictions and the distribution of predictions according to the explanatory sub-graph. This method achieves better performances from the point of view of the qualitative and quantitative similarity between the generated sub-graphs and the ground truth explanations.

Another application of DRL is to add or remove existing edges during adversarial attacks on GNNs [75, 76]. RLS2V [77] is a framework that uses DRL to learn structural changes in graphs, which is used to develop strategies for adversarial attacks on GNNs. Since GNNs are vulnerable to adversarial attacks that corrupt or poison the data used to train them. Q-learning and structure-to-vector-based attack methodology are learned to modify the graph structure. The purpose of DRL is to perform an attack aimed at evading detection during classification.

### 4.1 Multi-agent deep reinforcement learning

Multi-agent deep reinforcement learning needs coordination to efficiently solve certain works. Due to the size of joint action spaces, fully centralized control is often infeasible in these problems. The coordination graph-based method allows reasoning about the joint action based on the structure of interactions.

The coordination graph (CG) is introduced by Guestrin et al. [78], where a method for joint value estimation is presented that allows explicit modeling of the locality of interactions and formal reasoning about given joint actions. CG is a way to factorize a complex multi-agent Q-function. Rather than having a single joint Q function which would depend on the joint action of all agents, one could use a hypergraph to decompose this Q-function into a sum of Q functions across the edges, where each edge denotes a much lower dimensional Q function. Then finding the minimizing joint action can be done by passing messages along the edges of the coordination in a hypergraph.

MAGNet [79] represents policies for multi-agent environments based on relevant graphs and message-passing mechanisms. Here, the graphs are static and constructed based on heuristic rules. Multiple agents in the DGN model [80] are shown as nodes of a graph and relationships between them are learned as the observation encoder module in the environment. In the next step, by a convolutional kernel module, a multi-head point generation attention is defined to extract relational features between each agent and its neighbors in the local region. Q network module receives the extracted features of the former step to use them for determining the strategy which ultimately leads to cooperation between agents. In order to create an effective strategy in cooperation between agents, joint training between the encoder and Q network is done sequentially. This paper proves that GCN increases strongly cooperation among agents. This model is investigated in a grid-world platform MAgent.

Inspired by this idea [80], a model is presented in [81] that controls the connected autonomous vehicles (CAVs) as multi-agents by GNN and RL for cooperation between them. Information transfer for connected autonomous vehicles attains through the onboard sensors of nearby human-driven vehicles (HDVs) as local information and also from other connected autonomous vehicles the global information is obtained via connectivity channels. This information helps to define the graph structure. Within the local network, information passes from HDVs to CAVs. From the global network, all the CAVs can share knowledge including locally sensed information and their own information. Here, the environment contains a variable number of agents and makes a dynamic length output that matches with CAVs driving operations. Due to the variable number of agents, it is difficult to use joint training for each agent with its distinct Q network. Also, joint training is not scalable because by increasing the number of agents, the number of parameters for distinct Q networks will grow exponentially. One efficient method for solving these challenges is to apply a shared centralized Q network for all agents to determine their actions. Using the combination of GCN and deep Q network can have collaborative and safe controlling for lane-changing decisions in different traffic.

### 4.2 Multi-task deep reinforcement learning

MTDRL prepares a learning framework for coordinating and exploiting commonalities between multiple tasks in order to learn data efficiency, and robustness policies with improved efficiency, and generalization. Compatible state-action spaces are the main assumption in a MTDRL process such as the same dimensions of states and actions across multiple tasks. This issue is supported by GNNs due to capable of processing graphs with arbitrary sizes.

One of the applications of GNN in a MTDRL is in continuous control environments that use the features of each element of the MuJoCo agent to construct input graphs [82]. Each actuator has obtained the information from local sensors. A shared modular policy is defined as a global policy for each agent's actuators. Each limb of the MuJoCo agent is considered as a state with features containing positions, rotation, velocity, etc. that implements its independent policy to optimize joint reward function.

A framework in [83] is proposed to learn a job-shop scheduling problem (JSSP) by GNN and RL. The GNN section contains the creation of a graph from spatial features of the element of the job-shop problem and the RL section considers it as sequential decision-making by proximal policy optimization method (PPO) as a scheduling process.

## 5. Conclusion

In this survey, we tried to summarize about GNNs and RL and their relations. We had an overview of the challenges inherent in graph neural networks and multi-agent environments. Since, learning in collaborative multi-agent environments with dynamic, non-deterministic, and large state space has become a very important challenge in applications. Among these challenges, we can mention the effect of the size of the state space on the duration of learning, as well as the inefficient cooperation and the lack of proper coordination in decision-making between the agents. Also, when using reinforcement learning algorithms with the graph structure, the models will face challenges such as the difficulty of determining the appropriate learning goal and the long convergence time caused by trial and error-based learning. So, the integration of these methods leads to more realistic scenarios and more effective solutions to real-world problems. Researchers in this field have a significant impact on the progress of the combination of GNNs and DRL by providing newer models and architectures.

## Author details

Fatemeh Fathinezhad[1*], Peyman Adibi[1], Bijan Shoushtarian[1] and Jocelyn Chanussot[2]

1 Faculty of Computer Engineering, Artificial Intelligence Department, University of Isfahan, Isfahan, Iran

2 GIPSA-Lab, CNRS, Grenoble INP, University of Grenoble Alpes, Grenoble, France

*Address all correspondence to: fateme.fathinezhad@eng.ui.ac.ir

## References

[1] LeCun Y, Bengio Y, Hinton G. Deep learning. Nature. 2015;**521**(7553): 436-444

[2] Montavon G, Samek W, Müller KR. Methods for interpreting and understanding deep neural networks. Digital Signal Processing. 2018;**73**:1-15

[3] Zhou J, Cui G, Hu S, Zhang Z, Yang C, Liu Z, et al. Graph neural networks: A review of methods and applications. AI Open. 2020;**1**:57-81

[4] Wu Z, Pan S, Chen F, Long G, Zhang C, Philip SY. A comprehensive survey on graph neural networks. IEEE Transactions on Neural Networks and Learning Systems. 2020;**32**(1):4-24

[5] Eck D, Schmidhuber J. A first look at music composition using LSTM recurrent neural networks. Istituto Dalle Molle Di Studi Sull Intelligenza Artificiale. 2002;**103**:48

[6] Buşoniu L, Babuška R, De Schutter B. Multi-agent reinforcement learning: An overview. In: Innovations in Multi-Agent Systems and Applications-1, 2010th edition. Germany, Springer Berlin: Springer Verlag; August 11, 2010. pp. 183-221

[7] Mnih V, Kavukcuoglu K, Silver D, Rusu AA, Veness J, Bellemare MG, et al. Human-level control through deep reinforcement learning. Nature. 2015; **518**(7540):529-533

[8] Gilmer J, Schoenholz SS, Riley PF, Vinyals O, Dahl GE. Message passing neural networks. In: Machine Learning Meets Quantum Physics. Switzerland: Springer; 2020. pp. 199-214

[9] Hwang D, Yang S, Kwon Y, Lee KH, Lee G, Jo H, et al. Comprehensive study on molecular supervised learning with graph neural networks. Journal of Chemical Information and Modeling. 2020;**60**(12):5936-5945

[10] Bronstein MM, Bruna J, LeCun Y, Szlam A, Vandergheynst P. Geometric deep learning: Going beyond euclidean data. IEEE Signal Processing Magazine. 2017;**34**(4):18-42

[11] Hamilton WL, Ying R, Leskovec J. Representation learning on graphs: Methods and applications. IEEE Data Engineering Bulletin. arXiv: 1709.05584. 2017;**40**(3):52-74

[12] Ying R, He R, Chen K, Eksombatchai P, Hamilton WL, Leskovec J. Graph convolutional neural networks for web-scale recommender systems. In: Proceedings of the 24th ACM SIGKDD international conference on knowledge discovery and data mining, Washington DC. 2018

[13] Monti F, Frasca F, Eynard D, Mannion D, Bronstein MM. Fake news detection on social media using geometric deep learning. arXiv: 1902.06673. 2019

[14] Rossi E, Monti F, Bronstein MM, Liò P. NCRNA classification with graph convolutional networks. In: KDD Workshop on Deep Learning on Graphs. Anchorage, Alaska, USA: Association for Computing Machinery; 2019

[15] Zitnik M, Agrawal M, Leskovec J. Modeling polypharmacy side effects with graph convolutional networks. Bioinformatics. 2018;**34**(13): 457-466

[16] Veselkov K et al. Hyperfoods: Machine intelligent mapping of cancer-

beating molecules in foods. Scientific Reports. 2019;**9**(1):1-12

[17] Gainza P et al. Deciphering interaction fingerprints from protein molecular surfaces using geometric deep learning. Nature Methods. 2019;**17**: 184-192

[18] Estrach JB, Zaremba W, Szlam A, LeCun Y. Spectral networks and deep locally connected networks on graphs. In: 2nd International Conference on Learning Representations (ICLR), Banff, AB, Canada. Vol. 2014; 2014

[19] Kipf TN, Welling M. Semi-supervised classification with graph convolutional networks. In: International Conference on Learning Representations (ICLR), 5th ICLR (Poster), Toulon, France. 2017. Available from: OpenReview.net

[20] Tang S, Li B, Yu H. ChebNet: Efficient and stable constructions of deep neural networks with rectified power units using chebyshev approximations. arXiv preprint arXiv: 1911.05467. 2019

[21] Veličković P, Cucurull G, Casanova A, Romero A, Lio P, Bengio Y. Graph attention networks. arXiv preprint arXiv:1710.10903. 2017

[22] Hamilton W, Ying Z, Leskovec J. Inductive representation learning on large graphs. Advances in Neural Information Processing Systems. 2017; **30**:1025-1035

[23] Oh J, Cho K, Bruna J. Advancing graphsage with a data-driven node sampling. arXiv preprint arXiv: 1904.12935. 2019

[24] Pei Y, Yang J, Wang J, Xu P, Zhou T, Wu F. An emergency control strategy for undervoltage load shedding of power system: A graph deep reinforcement learning method. IET Generation, Transmission & Distribution. 2023;**17**: 2130-2141

[25] Zhang M, Chen Y. Link prediction based on graph neural networks. Advances in Neural Information Processing Systems. 2018; **31**:5171-5181

[26] Wu J, He J, Xu J. Net: Degree-specific graph neural networks for node and graph classification. In: Proceedings of the 25th ACM SIGKDD International Conference on Knowledge Discovery & Data Mining. Association for Computing Machinery, New York NY, United States. 2019. pp. 406-415

[27] Tsitsulin A, Palowitch J, Perozzi B, Müller E. Graph clustering with graph neural networks. arXiv preprint arXiv: 2006.16904. 2020

[28] Wang WY, Li J, He X. Deep reinforcement learning for NLP. In: Proceedings of the 56th Annual Meeting of the Association for Computational Linguistics: Tutorial Abstracts. Association for Computational Linguistics, Melbourne Convention and Exhibition Centre. 2018. pp. 19-21

[29] Haydari A, Yılmaz Y. Deep reinforcement learning for intelligent transportation systems: A survey. IEEE Transactions on Intelligent Transportation Systems. 2020;**23**(1): 11-32

[30] Hu YJ, Lin SJ. February. Deep reinforcement learning for optimizing finance portfolio management. In: 2019 Amity International Conference on Artificial Intelligence (AICAI), Amity University Dubai Campus Dubai International Academic City. Dubai: IEEE; 2019. pp. 14-20

[31] Coronato A, Naeem M, De Pietro G, Paragliola G. Reinforcement learning for intelligent healthcare applications: A survey. Artificial Intelligence in Medicine. 2020;**109**:101964

[32] Gu S, Holly E, Lillicrap T, Levine S. Deep reinforcement learning for robotic manipulation with asynchronous off-policy updates. In: 2017 IEEE International Conference on Robotics and Automation (ICRA). Marina Bay Sands, Singapore: IEEE; 2017. pp. 3389-3396

[33] Zhao X, Zhang L, Ding Z, Xia L, Tang J, Yin D. Recommendations with negative feedback via pairwise deep reinforcement learning. In: Proceedings of the 24th ACM SIGKDD International Conference on Knowledge Discovery & Data Mining, Association for Computing Machinery, New York NY, United States. 2018. pp. 1040-1048

[34] Mnih V, Kavukcuoglu K, Silver D, Graves A, Antonoglou I, Wierstra D, et al. Playing atari with deep reinforcement learning. arXiv preprint arXiv:1312.5602. 2013

[35] Zhu Y, Mottaghi R, Kolve E, Lim JJ, Gupta A, Fei-Fei L, et al. Target-driven visual navigation in indoor scenes using deep reinforcement learning. In: 2017 IEEE International Conference on Robotics and Automation (ICRA). Marina Bay Sands, Singapore: IEEE; 2017. pp. 3357-3364

[36] Chu T, Wang J, Codecà L, Li Z. Multi-agent deep reinforcement learning for large-scale traffic signal control. IEEE Transactions on Intelligent Transportation Systems. 2019;**21**(3): 1086-1095

[37] Soleymani F, Paquet E. Deep graph convolutional reinforcement learning for financial portfolio management–deeppocket. Expert Systems with Applications. 2021;**182**:115127

[38] Littman ML. Markov games as a framework for multi-agent reinforcement learning. In: Machine Learning Proceedings 1994. New Brunswick, New Jersey: Morgan Kaufmann; 1994. pp. 157-163

[39] Amato C, Dibangoye JS, Zilberstein S. Incremental policy generation for finite-horizon Dec-POMDPs. In: Proceedings of the Thirty-Second International Conference on Automated Planning and Scheduling. Palo Alto, California USA: AAAI Press; 2009

[40] Oliehoek FA, Spaan MT, Vlassis N. Optimal and approximate q-value functions for decentralized POMDPs. Journal of Artificial Intelligence Research. 2008;**32**:289-353

[41] Dibangoye JS, Amato C, Buffet O, Charpillet F. Optimally solving Dec-POMDPs as continuous-state MDPs. Journal of Artificial Intelligence Research. 2016;**55**:443-497

[42] Foerster JN, Farquhar G, Afouras T, Nardelli N, Whiteson S. Counterfactual multi-agent policy gradients. In: AAAI conference on artificial intelligence. New Orleans, Louisiana, USA: AAAI Press; 2018. pp. 2974-2982

[43] Omidshafiei S, Pazis J, Amato C, How JP, Vian J. Deep decentralized multi-task multi-agent reinforcement learning under partial observability. In: Proceedings of the 34th International Conference on Machine Learning: 70. Sydney, Australia: PMLR; 2017. pp. 2681-2690

[44] Papoudakis G, Christianos F, Rahman A, Albrecht SV. Dealing with

non-stationarity in multi-agent deep reinforcement learning. arXiv preprint arXiv:1906.04737. 2019

[45] Burkov A, Chaib-Draa B. Reducing the complexity of multiagent reinforcement learning. In: Proceedings of the 6th International Joint Conference on Autonomous Agents and Multiagent Systems. 2007. pp. 1-3

[46] Minsky M. Steps toward artificial intelligence. Proceedings of the IRE. IEEE. 1961;**49**(1):8-30

[47] Ding S, Du W, Ding L, Zhang J, Guo L, An B. Multiagent reinforcement learning with graphical mutual information maximization. In: IEEE Transactions on Neural Networks and Learning Systems. 2023

[48] Lee ES, Zhou L, Ribeiro A, Kumar V. Graph neural networks for decentralized multi-agent perimeter defense. Switzerland, Spain and china: Frontiers in Control Engineering. 2023;**4**:1

[49] Naderializadeh N, Hung FH, Soleyman S, Khosla D. Graph convolutional value decomposition in multi-agent reinforcement learning. arXiv preprint arXiv: 2010.04740. 2020

[50] Thekumparampil KK, Wang C, Oh S, Li L-J. Attention-based graph neural network for semi-supervised learning. arXiv preprint arXiv: 1803.03735. 2018

[51] Veličković P, Cucurull G, Casanova A, Romero A, Liò P, Bengio Y. Graph attention networks. In: International Conference on Learning Representations. 2018;**1710.10903**

[52] Devlin J, Chang M-W, Lee K, Toutanova K. Bert: Pre-training of deep bidirectional transformers for language understanding. In: Proceedings of the 2019 Conference of the North American Chapter of the Association for Computational Linguistics: Human Language Technologies. Vol. 1, (Long and Short Papers). Hyatt Regency in Minneapolis: Association for Computational Linguistics; 2019. pp. 4171-4186

[53] Iqbal S, Sha F. Actor-attention-critic for multi-agent reinforcement learning. In: International Conference on Machine Learning. Long Beach, California, USA: PMLR; 2019. pp. 2961-2970

[54] Kok JR, Vlassis N. Sparse cooperative Q-learning. In: Proceedings of the Twenty-First International Conference on Machine Learning, Banff Alberta, Canada. 2004. p. 61

[55] Li S, Gupta JK, Morales P, Allen R, Kochenderfer MJ. Deep implicit coordination graphs for multi-agent reinforcement learning. In: Proceedings of the 20th International Conference on Autonomous Agents and Multiagent Systems (AAMAS-2021), International Foundation for Autonomous Agents and Multiagent Systems (IFAAMAS), London. arXiv preprint arXiv: 2006.11438. 2020

[56] Zhou M, Chen Y, Wen Y, Yang Y, Su Y, Zhang W, et al. Factorized q-learning for large-scale multi-agent systems. In: Proceedings of the First International Conference on Distributed Artificial Intelligence (DAI'19). October 2019. pp. 1-7. [Article ID: 7]

[57] Sunehag P, Lever G, Gruslys A, Czarnecki WM, Zambaldi V, Jaderberg M, et al., 2017. Value-decomposition networks for cooperative multi-agent learning. In: AAMAS '18: Proceedings of the 17th International Conference on Autonomous Agents and MultiAgent Systems, Systems, Richland, SC,

Stockholm Sweden. arXiv preprint arXiv:1706.05296.

[58] Rashid T, Samvelyan M, Schroeder C, Farquhar G, Foerster J, Whiteson S. Qmix: Monotonic value function factorisation for deep multi-agent reinforcement learning. In: International Conference on Machine Learning. PMLR; 2018;**21**:7234-7284

[59] Lowe R, Wu YI, Tamar A, Harb J, Pieter Abbeel O, Mordatch I. Multi-agent actor-critic for mixed cooperative-competitive environments. Advances in Neural Information Processing Systems. Long Beach, CA, USA. 2017;**30**

[60] Foerster J, Farquhar G, Afouras T, Nardelli N, Whiteson S. Counterfactual multi-agent policy gradients. In: Proceedings of the AAAI Conference on Artificial Intelligence. 2018;**32**(1).

[61] Jiang J, Dun C, Huang T, Zongqing L. Graph convolutional reinforcement learning. In: International Conference on Learning Representations, Addis Ababa, Ethiopia. 2020. Available from: OpenReview.net

[62] Wang RE, Everett M, How JP. R-MADDPG for partially observable environments and limited communication. arXiv preprint arXiv: 2002.06684. 2020

[63] Li S, Wu Y, Cui X, Dong H, Fang F, Russell, S. Robust multi-agent reinforcement learning via minimax deep deterministic policy gradient. In: Proceedings of the AAAI Conference on Artificial Intelligence. 2019;**33**:4213-4220

[64] Yang Y, Luo R, Li M, Zhou M, Zhang W, Wang J. Mean field multi-agent reinforcement learning. In: International Conference on Machine Learning. PMLR; 2018. pp. 5571-5580

[65] Kim W, Cho M, Sung Y. Message-dropout: An efficient training method for multi-agent deep reinforcement learning. In: Proceedings of the AAAI Conference on Artificial Intelligence, Hilton Hawaiian Village, Honolulu, Hawaii, USA. 2019;**33**(1):6079-6086

[66] Almasan P, Suárez-Varela J, Rusek K, Barlet-Ros P, Cabellos-Aparicio A. Deep reinforcement learning meets graph neural networks: Exploring a routing optimization use case. Computer Communications. Netherlands: Elsevier; 2022;**196**:184-194

[67] Ma Q, Ge S, He D, Thaker D, Drori I. Combinatorial optimization by graph pointer networks and hierarchical reinforcement learning. arXiv preprint arXiv:1911.04936. 2019

[68] Wang Q, Tang C. Deep reinforcement learning for transportation network combinatorial optimization: A survey. Knowledge-Based Systems. 2021;**233**:107526

[69] Zheng P, Xia L, Li C, Li X, Liu B. Towards self-X cognitive manufacturing network: An industrial knowledge graph-based multi-agent reinforcement learning approach. Journal of Manufacturing Systems. 2021;**61**:16-26

[70] Vithayathil Varghese N, Mahmoud QH. A survey of multi-task deep reinforcement learning. Electronics. 2020;**9**(9):1363

[71] Elsken T, Metzen JH, Hutter F. Neural architecture search: A survey. The Journal of Machine Learning Research. 2019;**20**(1):1997-2017

[72] Gao Y, Yang H, Zhang P, Zhou C, Hu Y. Graphnas: Graph neural architecture search with reinforcement learning. arXiv preprint arXiv: 1904.09981. 2019

[73] Zhou K, Song Q, Huang X, Hu X. Auto-gnn: Neural architecture search of graph neural networks. Machine Learning and Artificial Intelligence, a section of the Journal Frontiers in Big Data. 2022;**5**. arXiv preprint arXiv: 1909.03184. 2019

[74] Shan C, Shen Y, Zhang Y, Li X, Li D. Reinforcement learning enhanced explainer for graph neural networks. Advances in Neural Information Processing Systems. 2021;**34**:22523-22533

[75] Tang X, Li Y, Sun Y, Yao H, Mitra P, Wang S. Transferring robustness for graph neural network against poisoning attacks. In: Proceedings of the 13th International Conference on Web Search and Data Mining. 2020. pp. 600-608

[76] Zügner D, Akbarnejad A, Günnemann S. Adversarial attacks on neural networks for graph data. In: Proceedings of the 24th ACM SIGKDD International Conference on Knowledge Discovery & Data Mining. Association for Computing Machinery, New York NY, United States. London, United Kingdom. 2018. pp. 2847-2856

[77] Dai H, Li H, Tian T, Huang X, Wang L, Zhu J, et al. Adversarial attack on graph structured data. In: International Conference on Machine Learning. PMLR, Stockholmsmässan, Stockholm Sweden; 2018. pp. 1115-1124

[78] Guestrin C, Koller D, Parr R. Multi-agent planning with factored MDPs. Advances in neural information processing systems. Vancouver, British Columbia, Canada. 2001:14

[79] Malysheva A, Sung TT, Sohn CB, Kudenko D, Shpilman A. Deep multi-agent reinforcement learning with relevance graphs. arXiv preprint arXiv: 1811.12557. 2018

[80] Jiang J, Dun C, Huang T, Lu Z. Graph convolutional reinforcement learning. arXiv preprint arXiv: 1810.09202. 2018

[81] Chen S, Dong J, Ha P, Li Y, Labi S. Graph neural network and reinforcement learning for multi-agent cooperative control of connected autonomous vehicles. Computer-Aided Civil and Infrastructure Engineering. 2021;**36**(7):838-857

[82] Huang W, Mordatch I, Pathak D. One policy to control them all: Shared modular policies for agent-agnostic control. In: International Conference on Machine Learning. PMLR; 2020. pp. 4455-4464

[83] Park J, Chun J, Kim SH, Kim Y, Park J. Learning to schedule job-shop problems: Representation and policy learning using graph neural network and reinforcement learning. International Journal of Production Research. United Kingdom. 2021;**59**(11):3360-3377

Section 2

# Applications of Deep Learning and Reinforcement Learning

Chapter 3

# IoT Device Identification Using Device Fingerprint and Deep Learning

*Prashant Baral, Ning Yang and Ning Weng*


## Abstract

The foundation of security in IoT devices lies in their identity. However, traditional identification parameters, such as MAC address, IP address, and IMEI, are vulnerable to sniffing and spoofing attacks. To address this issue, this paper proposes a novel approach using device fingerprinting and deep learning for device identification. Device fingerprinting is generated by analyzing inter-arrival time (IAT), round trip time (RTT), or IAT/RTT outliers of packets used for communication in networks. We trained deep learning models, namely convolutional neural network (CNN) and CNN + LSTM (long short-term memory), using device fingerprints generated from TCP, UDP, ICMP packet types, ICMP packet type, and their outliers. Our results show that the CNN model performs better than the CNN + LSTM model. Specifically, the CNN model achieves an accuracy of 0.97 using the IAT device fingerprint of ICMP packet type, and 0.9648 using the IAT outlier device fingerprint of ICMP packet type on a publicly available dataset from the crawdad repository.


**Keywords:** Internet of Things, deep learning, device identification, security, device fingerprinting

## 1. Introduction

IoT is used in varied industries including automobile, manufacturing, agriculture, and medicine, etc. With the increase in the usage of IoTs in varied fields, the data transfer between edge devices over the network has also increased. While IoT bridges the gap between the digital and physical world, compromised IoT devices can bring dangerous consequences. Wireless networks are more at risk than wired networks. Frames are encrypted in wireless communication, but the management and control frames are not encrypted as per IEEE 802.1 standards. This causes the wireless device identity prone to spoofing and denial of service attacks. Node forgery, once the adversaries get hold of the security credentials, can cause a major security threat.

Adversaries may use a compromised node to send incorrect data. For example, if an IoT device sending the temperature in the industry gets compromised, it will ruin the product, and the owner must bear a great loss. Many cryptography techniques, such as WEP and WPA, can be easily compromised. IP address, MAC address, or IMEI number could be used for device identification, but there are scenarios where these addresses got spoofing [1]. The gravity of the impact the breach in IoT has on varied fields is substantial, and we need to come up with an appropriate security mechanism to reduce the risk of data being compromised by IoT device forgery.

Different metrics can be used for device identification such as IP address, MAC address, IMEI address, and other network parameters such as transmission time, transmission rate, inter-arrival time, and medium access time. Comparisons of different metrics for device identification are in **Table 1**. The parameters, such as MAC address and IP address, are easier to spoof, so the study has been made on finding out the important parameters that can distinguish the devices. In [2], transmission time, transmission rate, inter-arrival time, and medium access time have been compared. IAT and transmission time outperform the other parameters in device identification.

In this paper, we worked on the deep learning approach for device identification. Device fingerprint is created from the parameters extracted during the communication of a device with router. This device fingerprint is used to train the deep learning model and device identification.

We fingerprint a device using IAT, RTT, and its outliers and feed them to deep learning models for device identification. These parameters are easier to extract and are not spoofed that easily after creating the device fingerprint with them. Timestamps (from which IAT and transmission time are extracted) are generated at the receiver side, which makes it harder to sniff and spoof. The adversaries need to change their own behavior to get a hand on these parameters. IAT and RTT are varied for different devices due to different CPU configuration and clock frequency. IAT and RTT depend on cache configurations, data cache, instruction cache, clock frequency, busses, and NIC card. These hardware configurations have an impact on the packet transfer rates. The attackers might try to emulate the signature using different techniques such as (1) introduce delays in packets, (2) change the data rate, and (3) make a customized operating system. Even while considering such techniques for an attack, an attacker is not successful in emulating the device. The attacker must consider a spoofing a signature along with hiding its original signature.

We use deep learning to extract knowledge from the data. It allows us to better understand the system model and simulate. CNN learns the semantic in the images and patterns in the image graph. Similarly, LSTM is recognized as a good algorithm for the classification of time series data. We use these two deep learning algorithms for the classification of devices. In earlier research, mathematical tools such as Mann-Whitney U-Test were used, but these algorithms require much time invested in

| Metrics/ Parameters | MAC address | IP address | Transmission time | Inter-arrival time |
|---|---|---|---|---|
| Spoofing | Easy | easy | difficult | difficult |
| Property | Hardware | network | hardware and software | hardware and software |
| Privacy concerns | Low | low | high | high |
| Header generation | sender wireless card | sender wireless card | receiver wireless card | receiver wireless card |

**Table 1.**
*Comparison of parameters for fingerprinting.*

preprocessing of data. The machine learning approach also requires us to prepare structured data before feeding it to ML algorithms.

Deep learning algorithms have an advantage over these consequences as it learns through the unstructured data while going through each layer in deep learning algorithms. The key factor for using deep learning is its time for making a prediction. Deep learning has its parameters calculated while training, which is why, when we provide our fingerprint of the device, the prediction is made quickly. This would take more time if we had used ML or any other mathematical tools.

We use the dataset generated from our own setup, as well as a publicly available dataset for training the model. We use TCP, UDP, ICMP packets, ICMP packets, and outliers of those packets for creating the device fingerprint. A new method [3] of device identification by collecting the information of the device to generate a fingerprint of the hardware, which can be used for device identification has been introduced. They use four different types of packets, namely probe request, ping, TCP, and UDP packets, to generate IAT graphs and have lower accuracy using CNN for classification.

In our work, we fingerprint two devices: Samsung A20 and Samsung J5 Prime. We plot IAT and RTT of the packets (probe request for IAT and ping for RTT) of each device and used those as datasets and feed them to deep learning models for device identification. We also use the publicly available dataset from the crawdad repository [4] introduced by Radhakrishnan et al. [5] to verify our results. This dataset provides the IAT information collected actively and passively from different wireless devices using wire side observations in a local network. They captured traffic from 30 different devices including iPads, iPhones, netbooks, Google phones, IP cameras, Kindles, and IP printers, etc., from various applications and protocols such as TCP, UDP, Skype, ICMP, SCP, and Iperf. Our main contribution consists of:

- use parameters extracted from wireless communication to create a unique signature/fingerprint of hardware.

- different parameters (IAT and RTT) for creating unique signatures, which are separately used for training deep learning models.

- compared how well deep learning algorithm was in classification using different metrics.

- considered outliers in IAT graph to observe if it does better classification.

The remainder of the paper is organized as follows. Section 2 briefly discusses related work. Section 3 describes the device fingerprinting, setup, methods for extracting data, and creating image graphs, and preparation of datasets are also discussed in this section. This dataset is fed to deep learning models described in Section 4 for classification of device. Experimental results are presented in Section 5, and the paper is concluded in Section 6.

## 2. Related work

The use of IP address, MAC address, and IMEI number for device identification brings significant risks of critical information, and the device itself is compromised. This alerts the researcher to produce a flexible and effective technique for device identification [1, 6, 7]. For example, a new stack [1] for the identity of IoT is proposed as it differs from the traditional identity of network devices and survey on attribute-based authentication for the identity of IoT devices.

Neumann et al. [2] surveys different features of the MAC layer such as transmission rate, transmission time, and inter-arrival time, and evaluated them on two criteria for effectiveness, fingerprint similarity at different time, and fingerprint dissimilarity of two different devices. In [2], authors use the IAT packets from wireless devices for creating digital fingerprints and created a histogram where each bin specifies the frequency of IATs in a specified range. Here, histogram is the fingerprint used for the classification of the the device and used to identify known and unknown devices from the database. The author tested the scenario where a malicious user tries to emulate the known device by introducing delay to the packets. The author concluded that different software and hardware make it difficult to emulate the hardware. In [2], authors use a passive approach for fingerprinting and, Radhakrishnan et al. [5] extended the work [2] using active approach for device fingerprinting. In the passive approach, we just observe the wireless communication to/from the device and use the important features of packets. Instead, in the active approach, we inject the signal to get a response from the device to get useful features. Sandhya et al. [8] used CNN but considered all types of packets flowing from devices to AP for device classification. This might be practical, but a lower accuracy of 86% may be problematic from a security point of view.

In [5], the author used a ping application to communicate between a device on campus. In [9], the author used IAT of probe request to fingerprint the device and used Mann-Whitney U-test for the analysis if two samples are of the same distribution. Miettinen [10] used 23 features such as ICMP, TCP, HTTP, and size from different layers (data link layer, transport layer, network layer, and application layer, etc.). The work collects these features of 23 for 12 packets and used a random forest algorithm for classification. The accuracy for 17 out of 27 was obtained 95% and 50% for the rest devices (10).

Robyns et al. [11] introduce the idea of noncooperative MAC layer fingerprinting, which does not require cooperation with the device as it uses some adversary nodes at the monitoring station to capture and monitor the bits of MAC frames without the user's permission. This hampers the privacy of the user but provides security from attacks from outside. The accuracy, when used for classification of 50 to 100 devices, was between 67–80%, but the accuracy decreases rapidly from 33–15% when device numbers were increased.

Kohno et al. [12] used the clock skew for fingerprinting devices. The work measured the timestamp by time difference of the time stamps using the traces from Tcpdump. The work considered the scenario where IP addresses were changed during data collection. Maurice [13] used a probe request and response for fingerprinting, but the results were not that promising for similar devices. Cunche et al. [14] used probe request from an AP and in response got the list of wireless networks. The work used this vulnerability to identify people from the list of networks connected. Francois et al. [15] made use of behavioral fingerprinting and automatically disconnects the device, which has suspicious activity and asks it to reconnect based on the behavioral fingerprint. Sun et al. [16] use the fingerprinting method for localization of devices connected to Wi-Fi AP indoor or outdoor.

Xu [1] studied the challenges and opportunities in digital fingerprinting for both wired and wireless devices. The author extracted the features from the physical and MAC layers such as clock skew, IAT, transmission time, SSID, and frequency. The work concluded IAT and transmission time as good parameters for device classification based on accuracy.

Kulin et al. [17] used different algorithms such as k-NN, decision tree, logistic regression, and neural networks for device classification using publicly available datasets. The performance of k-NN, decision tree, and logistic regression was good, but neural networks performed poorer than other classification algorithms with an average precision of 0.47 and recall value of 0.46. It is a common understanding that neural networks should perform better than others, but this was not the case in this work.

## 3. Device fingerprinting

We set up the devices in the lab for extracting the information about the devices. First, we set up Raspberry Pi as a router. Next, we use Samsung A20 and Samsung J5 Prime as an edge device (target IoT devices). Wireless communication between the edge devices and router was recorded. In the sniffing applications, Wireshark captures the packets incoming and outgoing on Raspberry Pi. These captured packets are used to calculate IATs/RTTs of packets and plot IAT, RTT, and IAT outlier graphs. These graphs are used as datasets to train and test the model. Python program is used to plot, label, and split the dataset. A split training set trains the deep learning model, and the testing dataset validates it. Our overall methodology is depicted in **Figure 1** and explained in detail in a subsection of this section and Section 4.

### 3.1 Our setup

Our setup has Raspberry Pi as a router and phones as the edge devices. Raspberry Pi (acts as a router) broadcasts an access point. The packets sent from edge devices are captured at the router side, which has a packet sniffing tool installed. Wireshark is installed in Raspberry Pi which inspects, deciphers, and keeps track of all incoming and outgoing packets to/from it. As there might be many packets coming to the router, we use the filter to find the required packet. We collected the data in two ways: 1. Probe request and response and 2. Ping request and response.

Probe requests are the packets broadcast by wireless devices, which consist of supported data rates and their capabilities. The access point receives these requests and responses with packets consisting of SSID, supported data rates, and encryption type, etc. We used a sniffing tool, Wireshark, to passively sniff the packets at the router level and use those packets for making IAT graphs.

Ping sends the ICMP echo request packet to any device on the network and waits for the response from the target device. In our setup, we ping the edge device, and the edge device responds to the router. This packet communication of ICMP is passively observed and recorded by Wireshark. This data is used for making RTT graphs.

### 3.2 Analysis of data and create image graph

The data collected by a sniffing tool and must be processed to obtain IAT and RTT. We obtain data using a snipping tool in Raspberry Pi. These data are timestamps of incoming and outgoing packets. We process timestamps to calculate the IAT and RTT of the packet.

After we obtain the value of IAT and RTT of packets, we write a Python program to plot the graph and download it. IAT and RTT graph is plotted as a line graph of 100

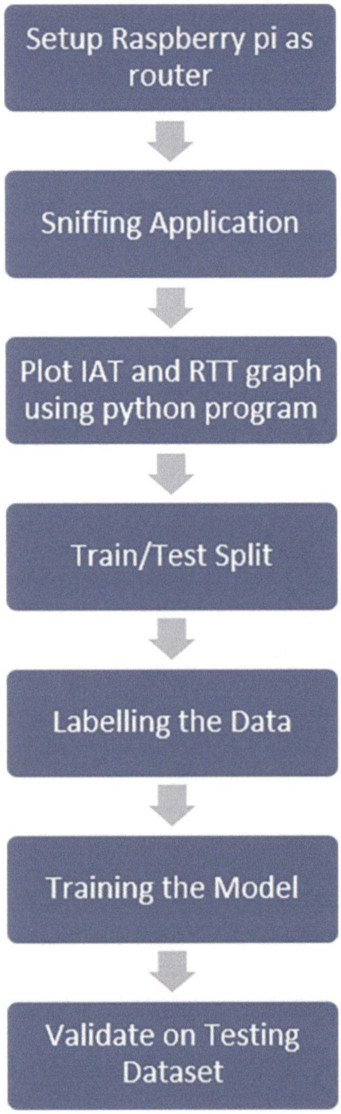


**Figure 1.**
*Methodology.*

IATs/RTTs. The plot of IAT and RTT is shown in **Figures 2–4**. We use IAT and RTT separately for device identification.

### 3.3 Preparation of data

The image obtained by plotting the graph must be labeled before we use that data for training and testing the model based on different metrics. We label the data using Python. For two phones, 0 represents Samsung A20 and 1 represents Samsung Prime. We split the total images into training data and testing data. For each IAT and RTT, we use 75 images for training and 30 for testing for each device (total 150 for training and 60 for testing). After creating an image and labeling it, we apply CNN and CNN + LSTM algorithms for image classification.

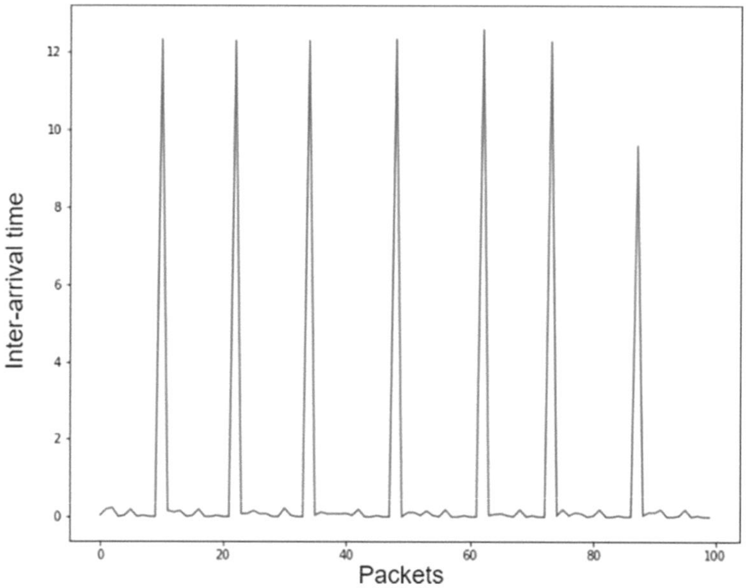


**Figure 2.**
*IAT graph from our setup.*

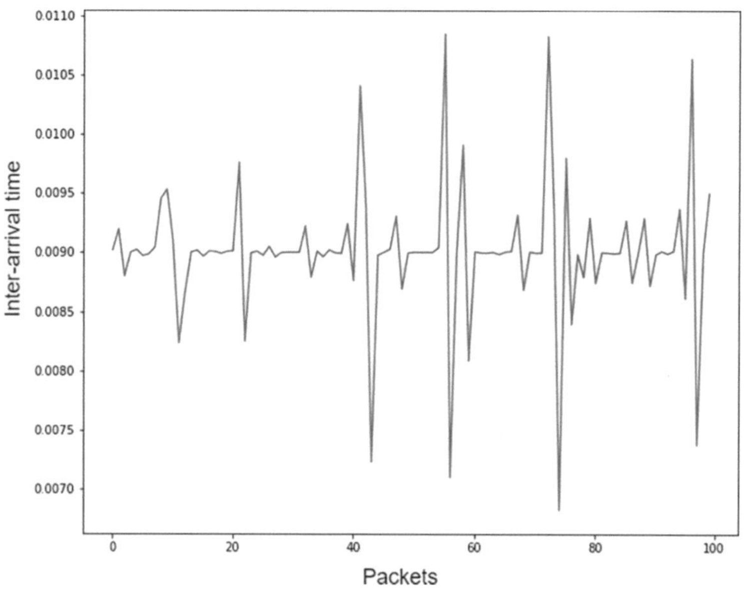


**Figure 3.**
*IAT graph from verification dataset.*

We use the dataset of IAT from crawdad, which was developed by Ulugac et al. [4]. The dataset is the collection of IATs of different devices. We use four devices: two iPad and two Dell notebooks for the verification of models. First, we use ICMP packets for generating the IAT graph. Since we are comparing the classification using a single packet type, multiple packet types, and an outlier, we also use TCP, UDP, and ICMP packets for generating the IAT graph and outliers. We plot a graph using 100

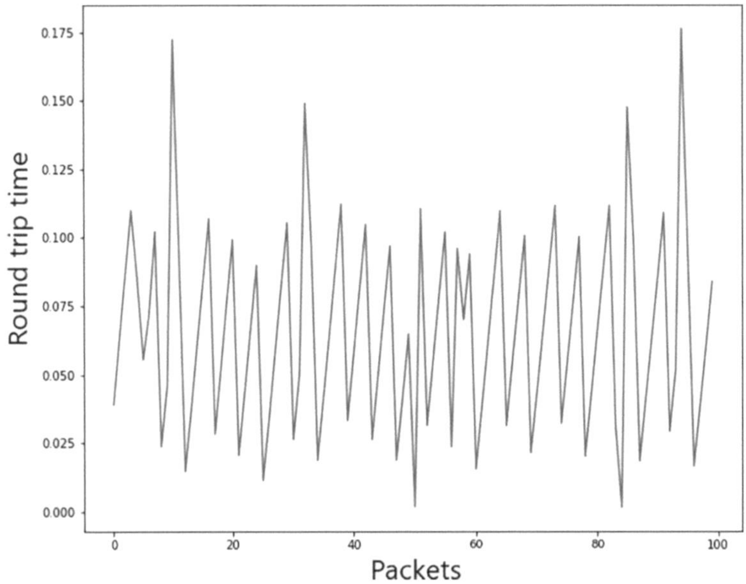


**Figure 4.**
*RTT graph from our setup.*

IATs. As in our setup, we similarly label zero for Dell notebook1, one for Dell notebook2, two for ipad1, and three for ipad2 and split it into a training and testing dataset.

## 4. Deep learning model for classification

We use a convolutional neural network (CNN) and a combination of a convolutional neural network and long short-term memory (CNN + LSTM) for device classification. Since we are using the image of time series data, we consider CNN due to its large breakthrough in image recognition. Moreover, CNN is very cost-effective due to the reduced number of parameters without losing the quality. Furthermore, due to the recognition of LSTM for time series data and our consideration of converting time series data to images and using the image for classification, we experiment if the combination of CNN + LSTM could give better results than CNN alone.

### 4.1 Convolutional neural network for device classification

The created image is colored, but for this classification problem, we convert the image into grayscale and reduce the image size to 256 * 256. Initially, it was 800 * 800. Then we split the labeled data into training and testing datasets and use the training set to train the CNN model. Our CNN model has the first convolution layer with 32 filters and a kernel size of 5 * 5. The input size of this layer was set to 256 * 256 * 1. Next, we use max-pooling with stride length 2; this helps in reducing the parameters by selecting the maximum from four (2 in x-direction and 2 in the y-direction). The next convolution layer in our model has 64 filters and a kernel size of 3 * 3. The input

| | | |
|---|---|---|
| INPUT LAYER | INPUT | ?,256,256 |
| | OUTPUT | ?.256.256 |
| CONV2D LAYER | INPUT | ?, 256,256 |
| | OUTPUT | ?. 256. 256. 32 |
| MAXPOOL2D LAYER | INPUT | ?,256,256,32 |
| | OUTPUT | ?,128,128,32 |
| CONV2D LAYER | INPUT | ?,128,128,32 |
| | OUTPUT | ?,126,126,48 |
| MAXPOOL2D LAYER | INPUT | ?,126,126,48 |
| | OUTPUT | ?,63,63,48 |
| CONV2D LAYER | INPUT | ?,63,63,48 |
| | OUTPUT | ?,62,62,64 |
| MAXPOOL2D LAYER | INPUT | ?,62,62,64 |
| | OUTPUT | ?,31,31,64 |
| FLATTEN LAYER | INPUT | ?,31,31,64 |
| | OUTPUT | ?.61504 |
| DENSE LAYER | INPUT | ?,61504 |
| | OUTPUT | ?.256 |
| DENSE LAYER | INPUT | ?,256 |
| | OUTPUT | ?.84 |
| DENSE LAYER | INPUT | ?,84 |
| | OUTPUT | ?.2 |

**Figure 5.**
*CNN model summary.*

to this layer is set by Keras. We again use max-pooling with stride length 2. The third convolution layer has 128 layers and a kernel size of 2 * 2, and we max-pooled with a stride length of 2 for this layer as well. For all these convolution layers, we use Rectified linear Unit (ReLU) as an activation function. Next, we use a flattened layer and two dense layers with 128 and 64 nodes followed by a dense layer with four nodes with softmax as activation function. **Figure 5** shows the model summary of CNN. The model is compiled using categorical cross-entropy for calculation of loss and Adam as the optimizer. We use both IAT and RTT data for training the CNN model and check how good was its classification using different metrics. Furthermore, we use an outlier of IAT data for classification. While training for different datasets, the number of nodes and epochs is changed.

### 4.2 Combination of CNN and LSTM for device classification

We combine CNN and LSTM using the concept of TimeDistributed layer. We provide n images at a time to the first TimeDistributed convolution layer; this applies

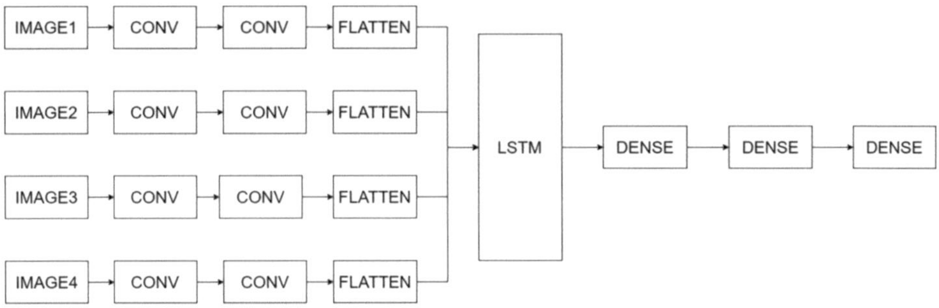


**Figure 6.**
*CNN + LSTM model.*

the same filter to the n images. We use the same three identical CNN layers but TimeDistributed. This is illustrated in **Figure 6**. The input to the first layer is n * 256 * 256 *1. Another input size is managed by Keras. This model has an additional LSTM layer with 32 nodes after CNN layers. The output of Maxpool2D is flattened to get one single vector. This is a feed to LSTM and a dense layer. **Figure 7** shows the model

| | | |
|---|---|---|
| INPUT LAYER | INPUT | ?,4,256,256 |
| | OUTPUT | ?,4,256,256 |
| TD CONV2D LAYER | INPUT | ?, 4,256,256 |
| | OUTPUT | ?,4, 256, 256, 32 |
| MAXPOOL2D LAYER | INPUT | ?,4,256,256,32 |
| | OUTPUT | ?,4,128,128,32 |
| TD CONV2D LAYER | INPUT | ?,4,128,128,32 |
| | OUTPUT | ?,4,127,127,64 |
| MAXPOOL2D LAYER | INPUT | ?, 4,127,127,64 |
| | OUTPUT | ?,4, 63,63,64 |
| TD CONV2D LAYER | INPUT | ?, 4,63,63,64 |
| | OUTPUT | ?,4, 63,63,128 |
| MAXPOOL2D LAYER | INPUT | ?, 4,63,63,128 |
| | OUTPUT | ?,4,31,31,128 |
| FLATTEN LAYER | INPUT | ?, 4,31,31,128 |
| | OUTPUT | ?,4,123008 |
| LSTM LAYER | INPUT | ?, 4,123008 |
| | OUTPUT | ?,4, 32 |
| DENSE LAYER | INPUT | ?, 4,32 |
| | OUTPUT | ?,4, 256 |
| DENSE LAYER | INPUT | ?, 4,256 |
| | OUTPUT | ?,4, 84 |
| DENSE LAYER | INPUT | ?, 4, 84 |
| | OUTPUT | ?,4, 2 |

**Figure 7.**
*CNN + LSTM model summary.*

summary of CNN + LSTM. LSTM makes use of chronological data and previous frame data to find what is useful in prediction. The model is compiled using categorical cross-entropy for calculation of loss and Adam as the optimizer. We use a combination of CNN and LSTM and observe how good the prediction the model can make. While training for different datasets, the number of nodes and epochs is changed.

### 4.3 Metrics for model evaluation

Evaluation of the model is an important task in data science. We need to make sure our model is not overfitted. Overfitting is a modeling error in statistics, which occurs due to the model aligning too closely to the limited data points. There are different techniques to prevent overfitting. Some of the techniques that we use are: reduce learning rate and dropout Layer. While training the model, we can monitor the validation accuracy and if it does not increase for a certain epoch, we reduce the learning rate by a certain factor. Below is the snippet of reducing learning rate where we monitor the validation loss and reduce the learning rate by a factor of 0.1 when for 3 consecutive epochs validation loss is increased.

*tf.Keras.callbacks.ReduceLROnPlateau(monitor = "val. loss," factor = 0.1,patience = 3,verbose = 0, andmin lr = 1e-6).*

Similarly, the dropout rate can be specified to the layer as the probability of setting each input to the layer to zero. Below is the code for adding the dropout layer. The rate is set to 0.3, which drops 0.3 of input units.

*model.add(Dense(128, activation = 'relu')).*
*model.add(Dropout(0.3)).*

The most common metric used for the evaluation of the algorithm is classification accuracy. Classification accuracy is equal to the number of correct predictions made divided by the total number of predictions made.

In our case, we use categorical cross-entropy for the calculation of loss, which makes the use of the probability of belonging to a class for the calculation of loss.

$$Classificationloss = -\sum_{i=1}^{outputsize} y_i logf(s)_i \qquad (1)$$

Where, $y_i$ is the class and $f(s)_i$ is the probability of belonging to that class. We also need to control the number of times we train the model. This is called epoch. Too much training can result in network overfitting to the training data. While training a model for certain epochs if validation error increases but the training loss decreases or remains constant, we can conclude that our model is overfitting as shown in **Figure 8**.

## 5. Results

Our setup has the phone Samsung A20, and Samsung Prime communicating with Raspberry Pi. As Section 3.2, we created the IAT graph using probe request and response from these devices to Raspberry Pi and prepared the data for feeding to CNN and evaluated the model. We trained the CNN model as in Section 4. A for 10 epochs and obtained the accuracy of 1.00 and loss of 0.0021 on training data. Accuracy in the validation dataset was 1.00 and loss of 0.0021. Using the IAT graph for classification and CNN + LSTM model and running for 30 epochs, the accuracy and loss were 1 and

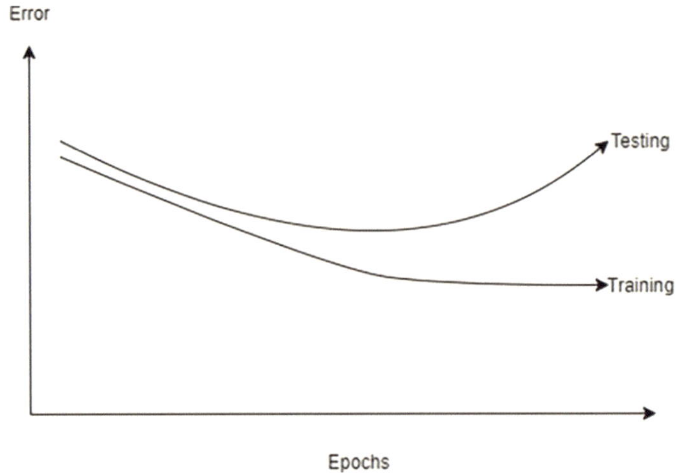


**Figure 8.**
*Model loss.*

0.0015 in the training dataset and 1 and 0.0011 in the validation dataset. Similarly, we created the RTT graph using ping as in Section 3.2 and trained for 10 epochs while feeding to CNN and 40 epochs while feeding to CNN + LSTM and achieved 100% accuracy in classification in both.

We used the dataset of IAT from crawdad, which was developed by Ulugac et al. [4] for verification. We used ICMP packets used by two Dell notebooks and two iPads communicating in the local area network. Using CNN for classification and running for 10 epochs, we achieved the accuracy of 1 and loss of 1.4 * 10–4 in the training dataset. We achieved an accuracy of 0.97 and a loss of 0.1326 in the validation dataset.

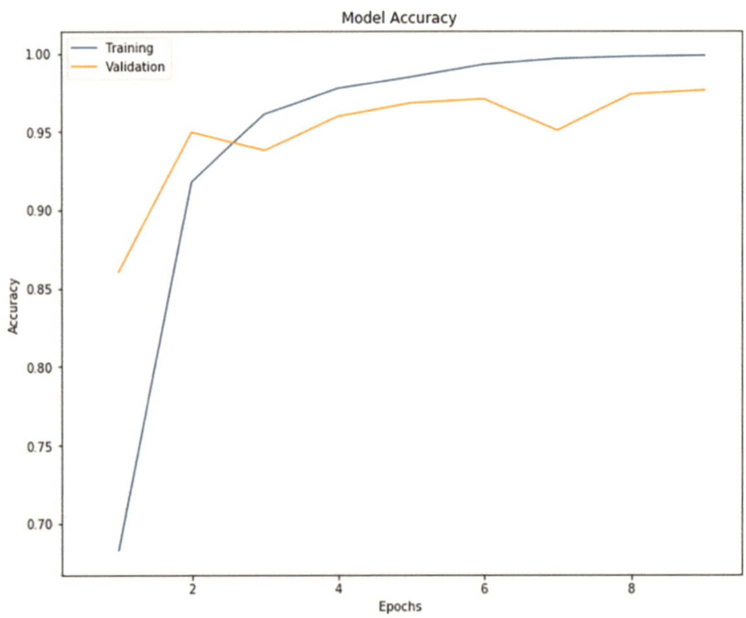


**Figure 9.**
*Accuracy using IAT(ICMP) as parameter from verification dataset using CNN.*

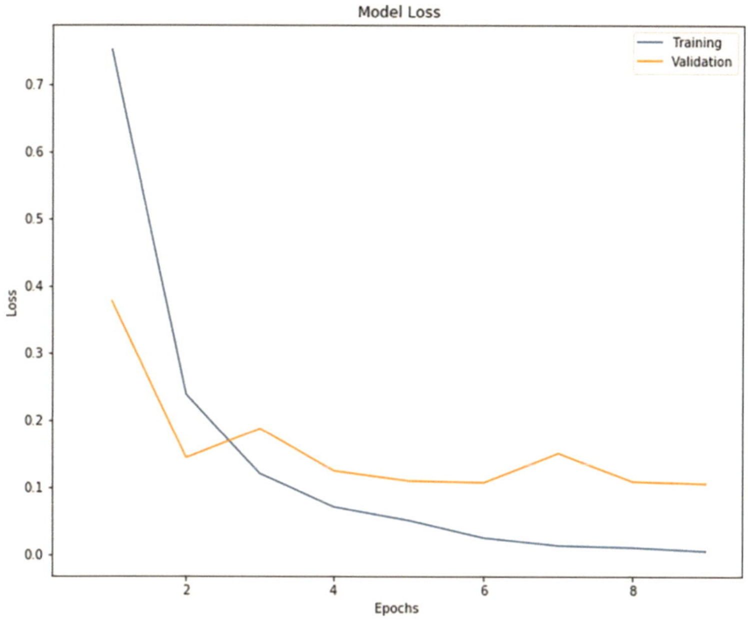


**Figure 10.**
*Loss using IAT(ICMP) as parameter from verification dataset using CNN.*

**Figures 9** and **10** show the learning curve of the CNN model. Using CNN + LSTM for classification and running for 35 epochs, we achieved an accuracy of 0.9463 and a loss of 0.1906 in the training dataset. We achieved an accuracy of 0.9060 and a loss of 0.3115 in the validation dataset. **Figures 11** and **12** show the learning curve of the CNN + LSTM model.

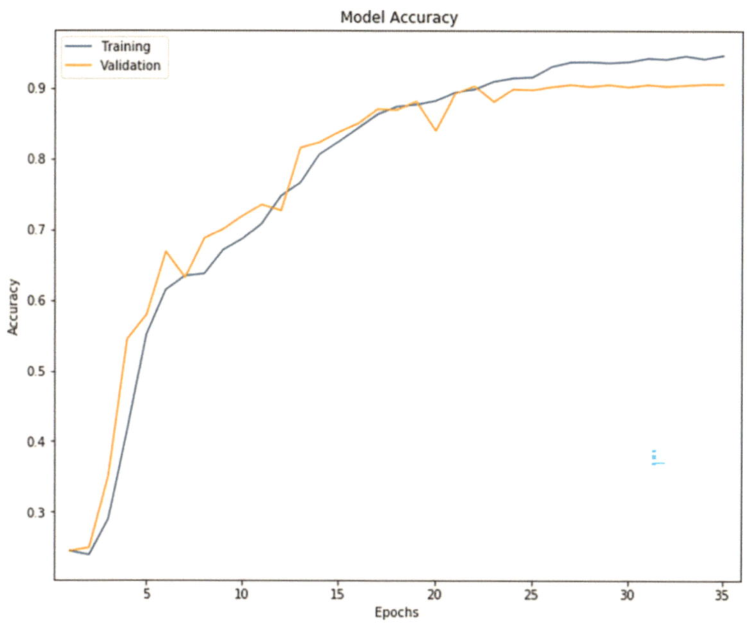


**Figure 11.**
*Accuracy using IAT(ICMP) as parameter from verification dataset using CNN + LSTM.*

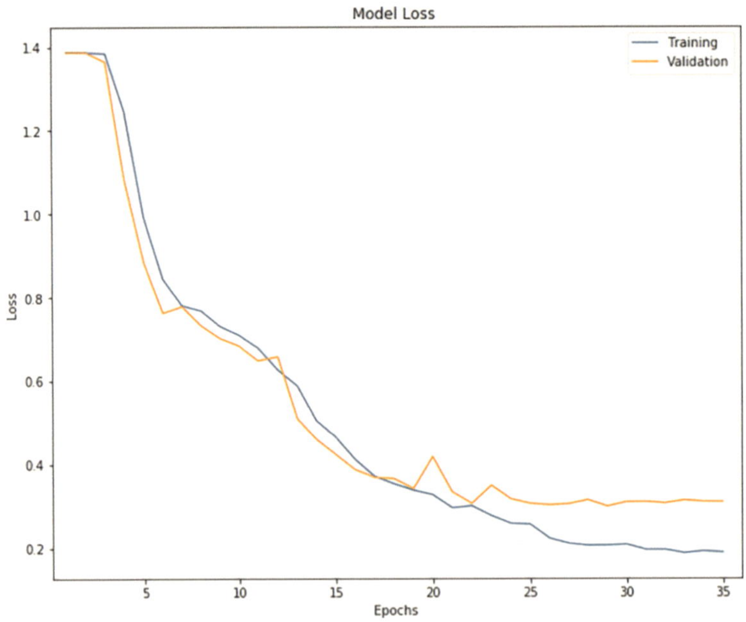


**Figure 12.**
*Loss using IAT(ICMP) as parameter from verification dataset using CNN + LSTM.*

After analyzing the IAT graph, we found that there is a regular pattern of outliers and considered if the outliers in the IAT graph can better classify a device using these deep learning algorithms. We utilized the outliers in the IATs of the verification dataset for four devices: two Dell notebooks and two iPads. There lies inter-burst latency between the IAT packets, and we utilize these for classification. We plotted the outlier graph for four devices considering their own threshold for each. We plotted outlier graphs and used CNN and CNN + LSTM algorithms for classification. We used the same CNN configurations ranging from convolution layers, input size, activation function, and number of layers, etc., for the classification using the IAT outlier graph. We ran the model for 10 epochs. We achieved the accuracy and loss of 0.9981 and 0.0079 and validation accuracy and loss of 0.9648 and 0.1397, respectively. **Figures 13** and **14** show the learning curve of the CNN model using an outlier dataset for training. We also used the same CNN + LSTM configurations ranging from convolution layers LSTM layer, activation function, and number of layers, etc., for the classification using the IAT outlier graph. We ran the model for 15 epochs. We achieved the accuracy and loss of 0.9870 and 0.0520 and validation accuracy and loss of 0.9574 and 0.1422, respectively. **Figures 15** and **16** show the learning curve of the CNN + LSTM model using an outlier dataset for training.

To validate the improvement of classification using single type packets (ICMP/probe request) in our work, we also classified the devices using TCP, UDP, and ICMP packet types from the same dataset of IAT from crawdad for classification as in [8]. The IAT graphs generated for these packet types were together used for training and testing the model. We trained the CNN model for 16 epochs and put the dropout layer after flattened layer to prevent overfitting. We used 18,000 training images and 6000 testing images and obtained an accuracy of 0.9656 and a loss of 0.0894; the validation accuracy and validation loss were 0.9290 and 0.3073, respectively. **Figures 17** and **18**

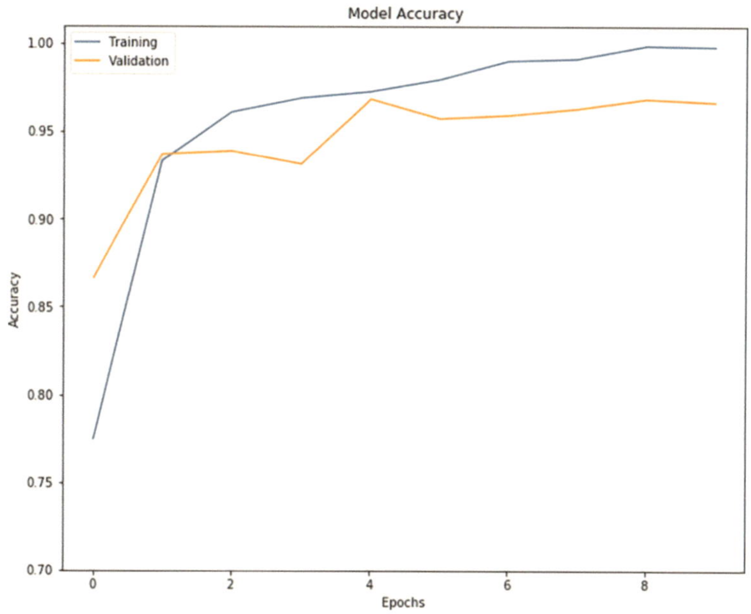


**Figure 13.**
*Accuracy using IAT(ICMP) outlier graph from verification dataset using CNN.*

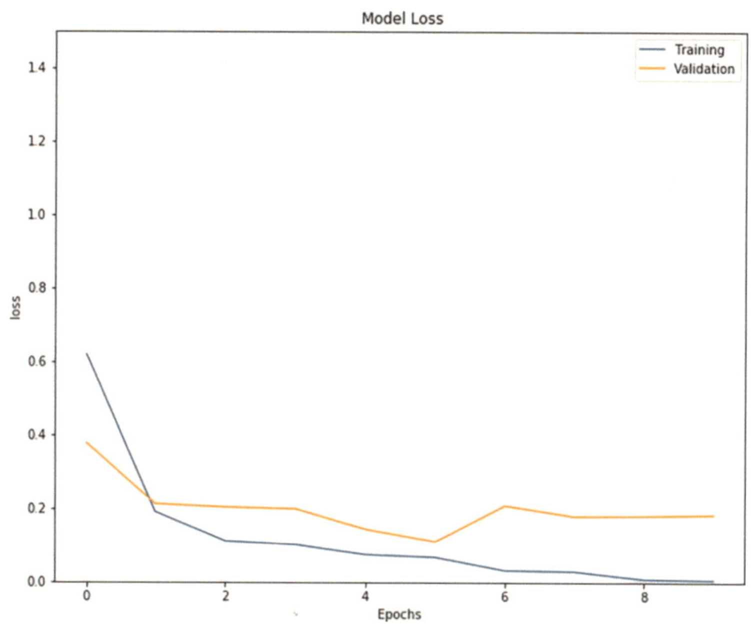


**Figure 14.**
*Loss using IAT(ICMP) outlier graph from verification dataset using CNN.*

show the learning curve of CNN model using image graphs of IAT generated using TCP, UDP, and ICMP packet types from the verification dataset.

Again, for this different type of packet, we considered the outliers and classified them using the outliers of IAT. We trained the CNN model for 20 epochs and put the

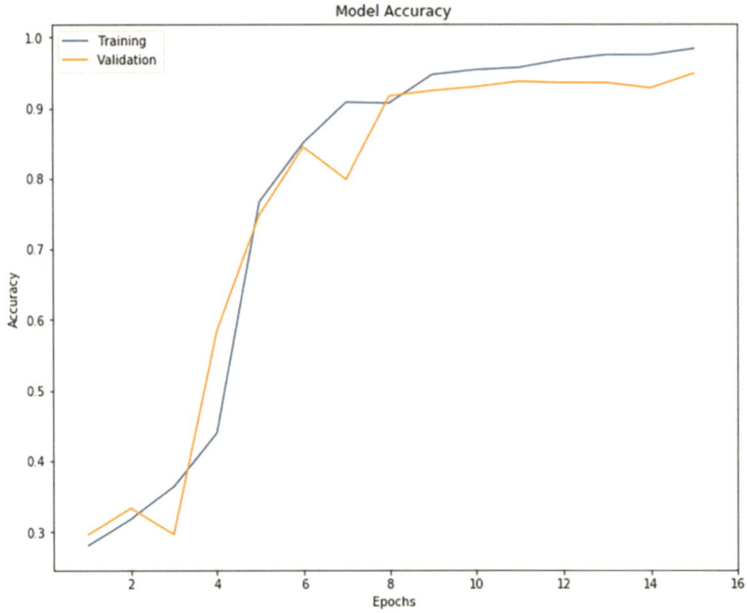


**Figure 15.**
*Accuracy using IAT (ICMP) outlier graph from verification dataset using CNN + LSTM.*

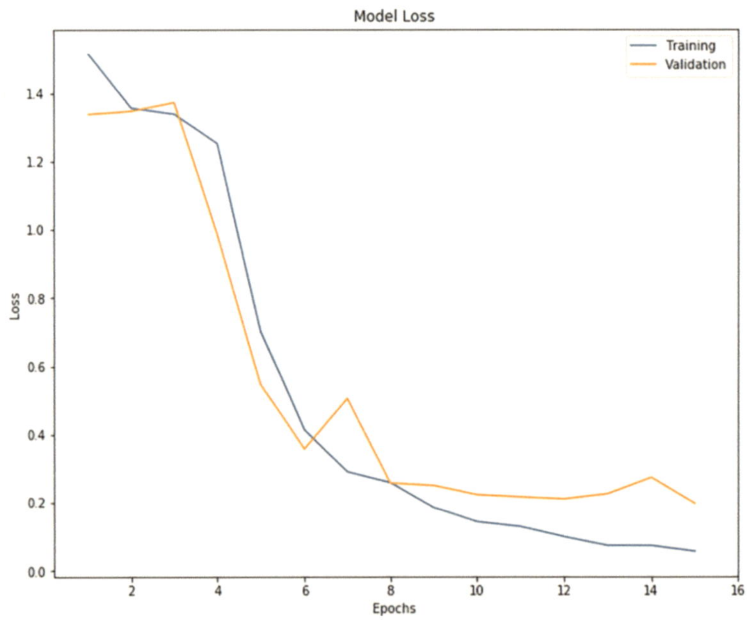


**Figure 16.**
*Loss using IAT (ICMP) outlier graph from verification dataset using CNN + LSTM.*

dropout layer after flatten layer to prevent overfitting. We used 5440 training images and 1700 testing images and obtained an accuracy of 0.8888 and a loss of 0.2704; the validation accuracy and validation loss were 0.8504 and 0.4344, respectively.

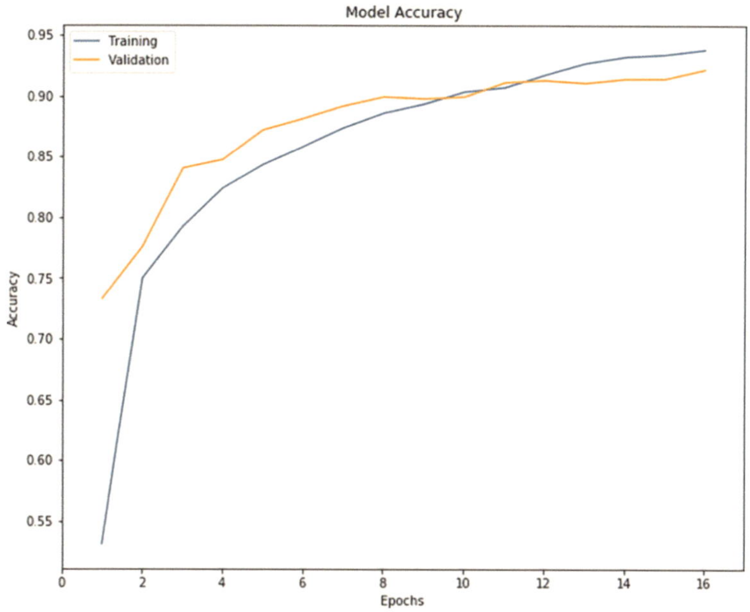


**Figure 17.**
*Accuracy using IAT (TCP, UDP, ICMP) as parameter from verification dataset using CNN.*

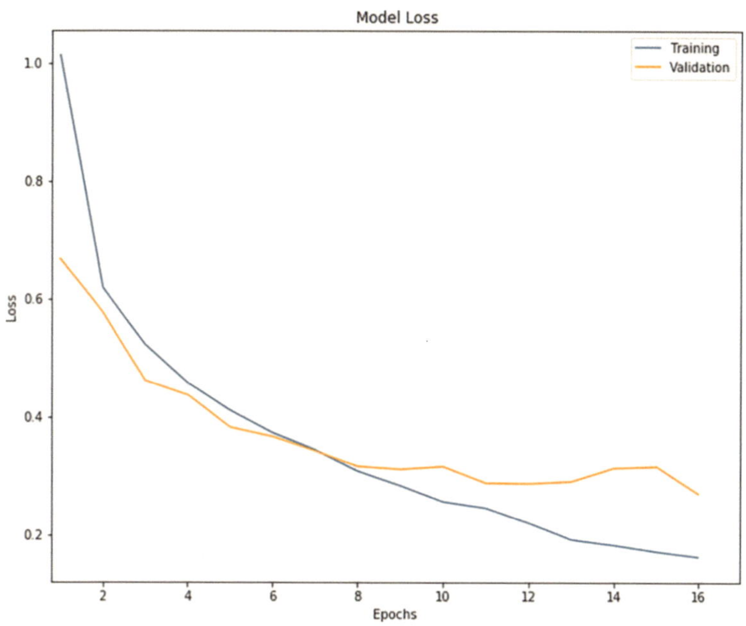


**Figure 18.**
*Loss using IAT (TCP, UDP, and ICMP) as parameter from verification dataset using CNN.*

**Figures 19** and **20** show the learning curve of the CNN model using image outlier graphs of IAT generated using TCP, UDP, and ICMP packet types from the verification dataset.

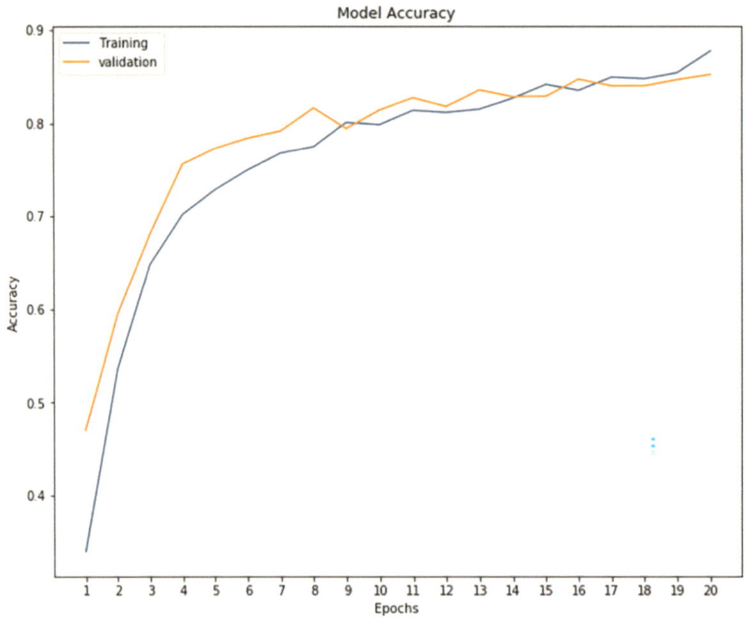


**Figure 19.**
*Accuracy using IAT(TCP, UDP, and ICMP) outlier graph from verification dataset using CNN.*

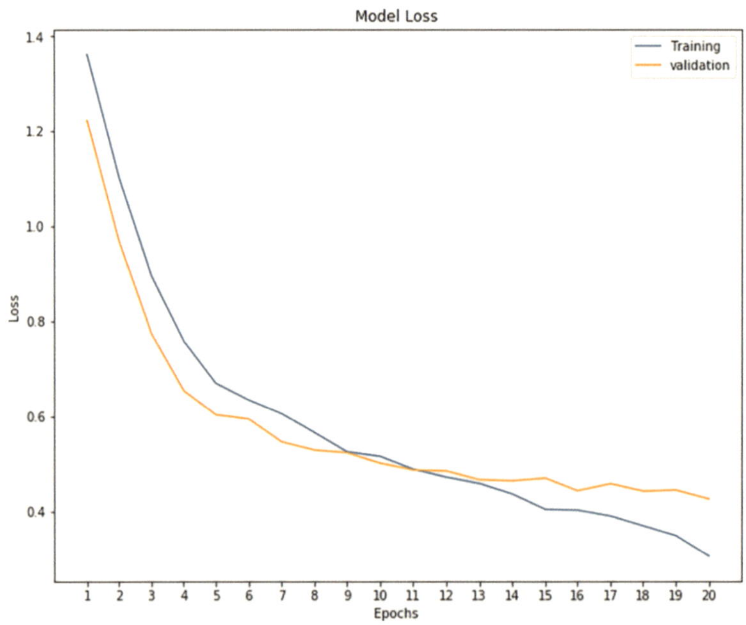


**Figure 20.**
*Loss using IAT(TCP, UDP, and ICMP) outlier graph from verification dataset using CNN.*

## 5.1 Comparison of models and parameters for IAT outlier graphs and IAT graphs from verification dataset

The summary of the model and parameters is shown in **Table 2**. When we used IAT graphs, the validation accuracy is 0.97 for CNN, which is better than

| Model/Parameters | IAT graphs (ICMP) | | IAT Outlier graphs (ICMP) | | IAT graphs (TCP, UDP, ICMP) | | IAT outlier graphs (TCP, UDP, ICMP) | |
|---|---|---|---|---|---|---|---|---|
| | val. Acc | val. Loss | val. Acc | val. Loss | val. Acc | val. Loss | val. Acc | val. Loss |
| CNN | 0.97 | 0.1326 | 0.9648 | 0.1397 | 0.9290 | 0.3073 | 0.8888 | 0.2704 |
| CNN + LSTM | 0.9060 | 0.5541 | 0.9574 | 0.1422 | — | — | 0.81 | 0.51 |

**Table 2.**
*Performance of models in terms of validation accuracy and validation loss using verification dataset.*

CNN + LSTM, in which case the validation accuracy is 0.9060. When we used the IAT outlier graph, the validation accuracy is 0.9648 for CNN and 0.9574 for CNN + LSTM. We observe that classification accuracy is similar in the case of CNN irrespective of the IAT graph or IAT outlier graph used in classification, but in the case of CNN + LSTM, the accuracy is lower, while using IAT graph for classification than IAT outlier graph.

We noticed that the results of the combination of CNN and LSTM cannot outperform the CNN alone model. The first reason is that the input of LSTM is a flattened version of CNN's output instead of a specific time series; therefore, the time dependence captured by LSTM may not reflect the relationship among input images. The second reason is that the used LSTM layer in the experiments has a small output size. In this case, some valuable information may be lost.

## 6. Conclusion

In this work, we classified devices using two parameters, namely inter-arrival time (IAT) and round-trip time (RTT), and two deep learning algorithms, namely CNN and a combination of CNN and LSTM. We used the IAT and RTT image graph as device fingerprint and model using two deep learning algorithms. We captured the packets using the packet snipping tool at Raspberry Pi(router) for two different setups. IAT and RTT were recorded for each device by snipping tool in real time. The security threat posed by adversaries once they forge the IoT device makes device identification a fundamental problem. The dynamic parameters that we used depend on hardware and software (CPU cache, data cache, and clock frequency, etc.), which makes it harder for intruders to create the fingerprint of a device. We used deep learning to extract the knowledge from data. The widespread recognition of CNN as a good algorithm for image classification encouraged us to use it. Moreover, as LSTM has made its name for the classification of time series data, we used a combination of CNN and LSTM because we were using an image graph of time series data for training the model. Our approach can be used to detect the malicious user if we store the fingerprint and match the fingerprint of the device trying to connect to the network before allowing it to connect. Our approach brings the alternative of using IMEI, IP and MAC address, cryptography security, and a digital certificate for device identification, which are prone to spoofing.

We used two different parameters and obtained good accuracy in our real setup. We also verified our model using the dataset available in public for a single ICMP packet and were able to achieve validation accuracy of 0.97 for CNN and 0.9060 for CNN + LSTM. We compared two deep learning algorithms for device identification.

Both models were good when we used a dataset that was generated from our setup, but while using the dataset from crawdad, CNN was more accurate in classification than CNN + LSTM. We further used IAT outlier graphs for classification and achieved validation accuracy of 0.9648 for CNN and 0.9574 for CNN + LSTM. To validate the improvement in classification accuracy using ICMP packet, we also classified the devices using TCP, UDP, and ICMP packet types from the verification dataset. We achieved good accuracy in using a single ICMP packet type for classification.

We collected RTT data in our setup and achieved good accuracy in classification. In the future, we can collect RTT data in a real scenario with many devices and use it for classification.

## Acknowledgements

This work is supported in part by the US National Science Foundation under Grant CC-2018919. Beside NSF grant support, Dr. Yang's work is also supported in part by the new hire startup fund from Southern Illinois University Carbondale.

## Conflict of interests

The authors declare that there are no conflicts of interest regarding the publication of this article.

## Author details

Prashant Baral[1†], Ning Yang[2] and Ning Weng[3*]

1 Advanced Micro Devices, Inc., Austin, TX, USA

2 Information Technology Program in the School of Computing, Southern Illinois University Carbondale, IL, USA

3 School of Electrical, Computer, and Biomedical Engineering, Southern Illinois University Carbondale, IL, USA

*Address all correspondence to: nweng@siu.edu

† These authors contributed equally.

## References

[1] Xu Q, Zheng R, Saad W, Han Z. Device fingerprinting in wireless networks: Challenges and opportunities. IEEE Communications Surveys & Tutorials. 2015;**18**(1):94-104

[2] Neumann C, Heen O, Onno S. An empirical study of passive 802.11 device fingerprinting. In: 2012 32nd International Conference on Distributed Computing Systems Workshops. Macau, China: IEEE; 2012. pp. 593-602

[3] Bratus S, Cornelius C, Kotz D, Peebles D. Active behavioral fingerprinting of wireless devices. In: Proceedings of the First ACM Conference on Wireless Network Security. New York, NY, USA: ACM; 2008. pp. 56-61

[4] Uluagac AS. "A. selcuk uluagac, crawdad dataset gatech/fingerprinting (v. 2014-06-09). 2014. Available from: https://crawdad.org/gatech/fingerprinting/20140609.

[5] Uluagac AS, Radhakrishnan SV, Corbett C, Baca A, Beyah R. A passive technique for fingerprinting wireless devices with wired-side observations. In: 2013 IEEE Conference on Communications and Network Security (CNS). Washington, D.C., USA: IEEE; 2013. pp. 305-313

[6] Hamad SA, Zhang WE, Sheng QZ, Nepal S. Iot device identification via network-flow based fingerprinting and learning. In: 2019 18th IEEE International Conference on Trust, Security and Privacy In Computing and Communications/13th IEEE International Conference on Big Data Science and Engineering (TrustCom/BigDataSE). Rotorua, New Zealand: IEEE; 2019. pp. 103-111

[7] Mazhar N, Salleh R, Zeeshan M, Hameed MM. Role of device identification and manufacturer usage description in iot security: A survey. IEEE Access. 2021;**9**:41 757-41 786

[8] Aneja S, Aneja N, Islam MS. Iot device fingerprint using deep learning. In: 2018 IEEE International Conference on Internet of Things and Intelligence System (IOTAIS). Bali, Indonesia: IEEE; 2018. pp. 174-179

[9] Desmond LCC, Yuan CC, Pheng TC, Lee RS. Identifying unique devices through wireless fingerprinting. In: Proceedings of the First ACM Conference on Wireless Network Security. New York, NY, USA: ACM; 2008. pp. 46-55

[10] Miettinen M, Marchal S, Hafeez I, Asokan N, Sadeghi A-R, Tarkoma S. Iot sentinel: Automated device-type identification for security enforcement in iot. In: 2017 IEEE 37th International Conference on Distributed Computing Systems (ICDCS). Atlanta, USA: IEEE; 2017. pp. 2177-2184

[11] Robyns P, Bonné B, Quax P, Lamotte W. Noncooperative 802.11 mac layer fingerprinting and tracking of mobile devices. Security and Communication Networks. 2017;**2017**: 1-21

[12] Kohno T, Broido A, Claffy KC. Remote physical device fingerprinting. IEEE Transactions on Dependable and Secure Computing. 2005;**2**(2):93-108

[13] Maurice C, Onno S, Neumann C, Heen O, Francillon A. Improving 802.11 fingerprinting of similar devices by cooperative fingerprinting. In: 2013 International Conference on Security and Cryptography (SECRYPT). Reykjavik, Iceland: IEEE; 2013. pp. 1-8

[14] Cunche M. I know your mac address: Targeted tracking of individual using wi-fi. Journal of Computer Virology and Hacking Techniques. 2014;**10**(4):219-227

[15] François J, State R, Engel T, Festor O. Enforcing security with behavioral fingerprinting. In: 2011 7th International Conference on Network and Service Management. Paris, France: IEEE; 2011. pp. 1-9

[16] Sun L, Chen S, Zheng Z, Xu L. Mobile device passive localization based on ieee 802.11 probe request frames. Mobile Information Systems. 2017;**2017**: 1-10

[17] Kulin M, Fortuna C, De Poorter E, Deschrijver D, Moerman I. Data-driven design of intelligent wireless networks: An overview and tutorial. Sensors. 2016; **16**(6):790

Chapter 4

# MultiRes Attention Deep Learning Approach for Abdominal Fat Compartment Segmentation and Quantification

*Bhanu K.N. Prakash, Arvind Channarayapatna Srinivasa, Ling Yun Yeow, Wen Xiang Chen, Audrey Jing Ping Yeo, Wee Shiong Lim and Cher Heng Tan*

## Abstract

Global increase in obesity has led to alarming rise in co-morbidities leading to deteriorated quality of life. Obesity phenotyping benefits profiling and management of the condition but warrants accurate quantification of fat compartments. Manual quantification MR scans are time consuming and laborious. Hence, many studies rely on semi/automatic methods for quantification of abdominal fat compartments. We propose a MultiRes-Attention U-Net with hybrid loss function for segmentation of different abdominal fata compartments namely (i) Superficial subcutaneous adipose tissue (SSAT), (ii) Deep subcutaneous adipose tissue (DSAT), and (iii) Visceral adipose tissue (VAT) using abdominal MR scans. MultiRes block, ResAtt-Path, and attention gates can handle shape, scale, and heterogeneity in the data. Dataset involved MR scans from 190 community-dwelling older adults (mainly Chinese, 69.5% females) with mean age—67.85 $\pm$ 7.90 years), BMI 23.75 $\pm$ 3.65 kg/m$^2$. Twenty-six datasets were manually segmented to generate the ground truth. Data augmentations were performed using MR data acquisition variations. Training and validation were performed on 105 datasets, while testing was conducted on 25 datasets. Median Dice scores were 0.97 for SSAT & DSAT and 0.96 for VAT, and mean Hausdorff distance was <5 mm for all the three fat compartments. Further, MultiRes-Attention U-Net was tested on a new 190 datasets (unseen during training; upper & lower abdomen scans with different resolution), which yielded accurate results. MultiRes-Attention U-Net significantly improved the performance over MultiResUNet, showed excellent generalization and holds promise for body-profiling in large cohort studies.

**Keywords:** MultiRes attention, deep learning, fat compartments, abdomen, subcutaneous fat compartments, visceral fat

## 1. Introduction

Obesity is a globally growing epidemic which has affected more than 2 billion adults, and many teens (18 years plus) are overweight, of which 650 million are obese [1]. Anthropometric measurements, waist-to-hip ratio, body mass index (BMI), waist circumference, does not explicitly distinguish fat mass, and quantity of fat present in visceral, and subcutaneous compartments. Literature, highlights that accumulation of fat leads to insulin resistance, oncologic and cardiovascular diseases [2–4] affecting the quality of life. Hence, body composition analysis to determine the amount of adipose and muscle tissue is of medical importance for obesity risk analysis. Magnetic resonance imaging (MRI) and computed tomography (CT) can characterize fat and non-fat tissues [5]. Among the imaging modalities, MR is more efficient in tissue characterization compared to CT for quantification of body fat volume [6, 7]. By quantifying different fat compartments from the imaging scans, we can perform body composition analysis. Manual quantification of fat and muscle volumes from the imaging scans is tedious and time-consuming, leading to loss clinical man-hours.

Anatomically, the subcutaneous adipose tissue compartments (superficial: SSAT and deep: DSAT) are separated by thin fascia, whereas the visceral adipose tissue (VAT) is found in-between internal and external abdominal boundaries. VAT is around the internal organs and discontinuous whereas SAT (SSAT+DSAT) is continuous. Fat depots are irregular in shape, lack texture, and vary across abdominal profile as demonstrated in **Figure 1** making it a challenging medical image segmentation task. Several semi-automated methodologies have been developed to reduce time and reduce bias [8–12]. These methodologies are less reliable and offer low accuracy as they depend on expert knowledge for fine-tuning image parameters.

Deep learning for image segmentation [13] has found many applications in medical image analysis and one such application is abdominal fat compartment segmentation. Several fat quantification studies use single contrast DIXON MR scan and 2D/3D U-Net architecture [14, 15] for SAT and VAT segmentation. Enhancement versions of Standard U-Net such as Competitive Dense Fully Convolutional Network (CDFNet), nnUNet, and Dense Convolutional Network (DCNet), which can handle complex image features, have been used for adipose tissue segmentation [16–18]. Attention gate model [AG] in 2D and 3D U-Net [19] has gained popularity in adipose tissue segmentation task as AG focuses on target structures of varying shapes and sizes by suppressing irrelevant regions and highlighting useful salient features [20, 21]. Ibtehaz et al. proposed a MultiRes block to address multiscale issues and ResPath to

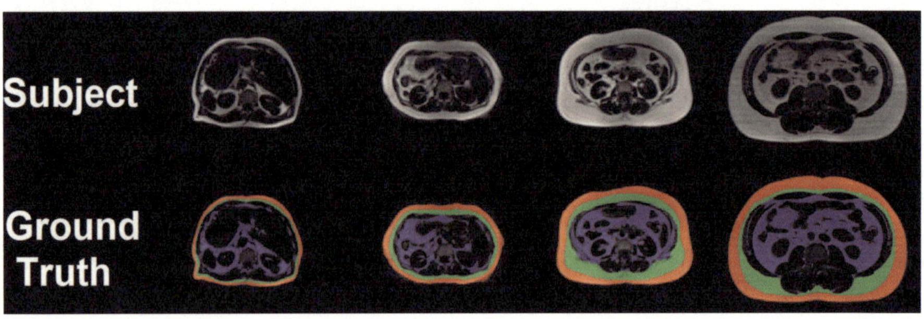


**Figure 1**
*Illustration of fat depots of SSAT (red), DSAT (green), and VAT (blue) varying shape, size across the abdominal profile.*

reduce adverse learning of features which might lead to a false prediction by skip connection of U-Net [22].

### 1.1 Study proposition

In our previous work on adipose fat depot segmentation, we had proposed patch-based 3D-ResUNet Attention [23] for fat depot segmentation, The patch-based framework failed to handle (i) different body compositions like lean, and moderately obese due to fixed patch sizes, and (ii) generalize to unseen abdominal region segmentation due to cataphoric forgetting of network, anatomical differences, and class imbalance. **Figure 2** illustrates a few failed cases from our previous work. Hence to overcome these drawbacks, we focused on the enhancement of MultiResUNet [23] by proposing a MultiRes-Attention U-Net architecture, with

i. a hybrid loss function to handle class imbalance, and

ii. attention gates for focused learning and improved prediction accuracy.

In this study, we also compare the performance of the proposed architecture against standard U-Net and MultiResUNet.

## 2. Materials and methods

### 2.1 MR data acquisition

Data sets of 190 elderly Asians (aged >50 years, residing within the community) who participated in characterization of early sarcopenia to assess functional decline

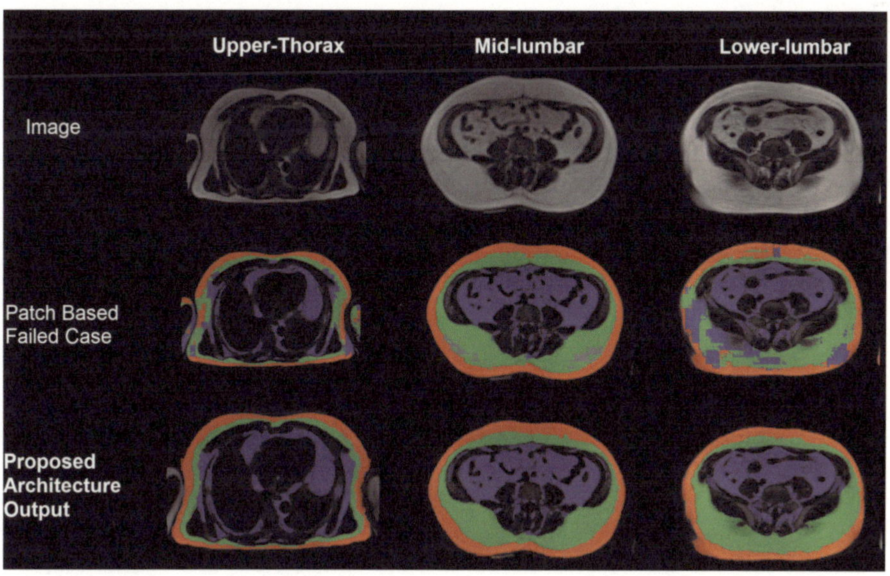


**Figure 2**
*Illustration of failed cases of our previous work on patch-based 3D-ResUNet attention vs. proposed architecture.*

study was used in our study [24]. The MR abdominal scans were acquired using a 3D modified breath-hold T1-weighted Dixon sequence. Subjects were advised a 20 s breath hold during the scans. The scans were performed on a 3T Siemens Magnetom Trio MRI scanner with TR/TE/FA/Bandwidth: 6.62 ms, 1.225 ms, 100, and 849 Hz/pixel, respectively. The study group consisted of mainly Chinese (91.6%) ethnicity having mean age was 67.85 ± 7.90 years, BMI 23.75 ± 3.65 kg/$m^2$, and predominantly female (69.5%) subjects. As the study subjects were elderly, many had common comorbidities such as hypertension, diabetics, and hyperlipidemia. National Healthcare board reviewed the cohort study with written consent from all subjects.

Data set can be considered as heterogeneous as it included (i) subjects from different ages (ii) scans covering different anatomical regions—thoracic, lumbar, and sacral (iii) variations in fat accumulation in different compartments based on body composition and (iv) acquisitional variations like—image dimensions, slice thickness, breathing/motion artifacts, etc.

Manual (radiology experts) ground truths were generated in 26 data sets out of 190 scans covering L1-L5 regions. The data with ground truths were subjected to MR-acquisition based data augmentation to scale the number from 26 to 130 to create training data sets.

### 2.2 Fat segmentation

A 3-stage segmentation framework was envisaged to quantify abdominal fat depots (i) Preprocessing stage which included (a) arm region removal, (b) data augmentation to increase the number of data sets, and (c) conversion of 3D MR images into 2D slices; (ii) Segmentation stage—"MultiRes-Attention U-Net" architecture for segmentation of abdominal regions into SSAT/DSAT/VAT (three class) regions and (iii) Postprocessing stage—image reconstruction 2D to 3D and fat depot quantification.

### 2.3 Preprocessing

All the training/testing data were subjected to quality check to assess motion artifacts originating from breathing, and fat-water swaps. Auto-check was developed to ensure training dataset slices match with the marked ground-truth slices. Arm region artifacts were removed automatically using the projection method [21]. Four different data augmentations were performed once before training these included (i) Random Noise (ii) Random Ghosting (iii) Random Bias Field (iv) Blur augmentation [23] to increase the total number of datasets. Finally, 3D MR scans were converted to 2D slices for training/testing the proposed deep learning architecture.

### 2.4 MultiRes-attention U-Net

In standard deeper convolutional network, input data goes through multiple convolutions to obtain salient spatial features leading to vanishing gradient problem. The architectures like ResNet [25] adopt summation of connect of all preceding feature maps leading to high memory demanding network. DenseNet [26] introduces "dense connections", where each layer in the network is connected to every other layer, instead of only connected to previous layers as in standard network architecture but fail to handle multi-scale issue. To handle multi-scale issue of fat depots which vary in shape, size, and improve semantic segmentation which is memory efficient.

We proposed MultiRes-Attention U-Net which is a modified version of MultiResUNet with attention which contains (i) MultiRes block, (ii) ResAtt-Path, and (iii) Attention gate model.

## 2.5 MultiRes block

Two sequential convolutional layers at each level in U-Net [24] are substituted with a proposed MultiRes block (similar to dense block in denseNet [26]) with the residual path, (as in ResNet [25]) as shown in **Figure 3**. multiRes block contains Inception-like modules with parallel convolution filters of 3×3, 5×5, and 7×7 to capture spatial features from different scales. However, they are not memory efficient. To reduce the memory, we factorized a large filter into a sequence of 3×3 filters with a gradual increase in the number of filters at each layer as shown in **Figure 3**.

## 2.6 ResAtt-path

Skip connections of standard U-Net are modified as ResAtt-Path by including non-linear convolution filters of 3×3 and a residual path with 1×1 filters. The number of convolution filters (3×3) reduces in each level of the encoding section of U-Net as shown in **Figure 4**. These ResAtt-Path overcomes the drawback of U-Net short connections by merging of low and high levels features at the decoder.

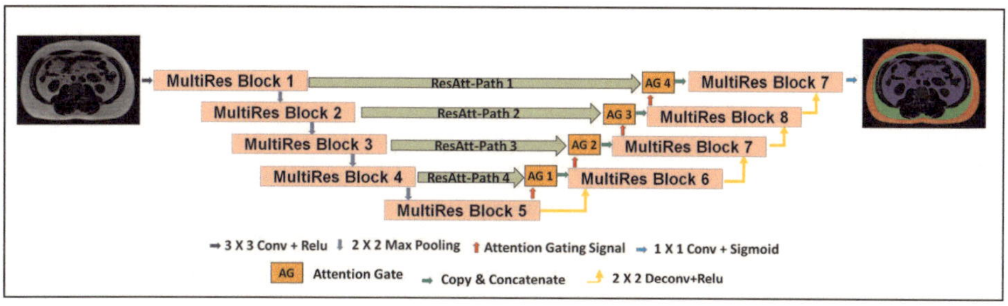


**Figure 3.**
*Proposed MultiRes-attention U-Net architecture with MultiRes Block, ResAtt-path and attention gate block at the decoder to aggregate attention features.*

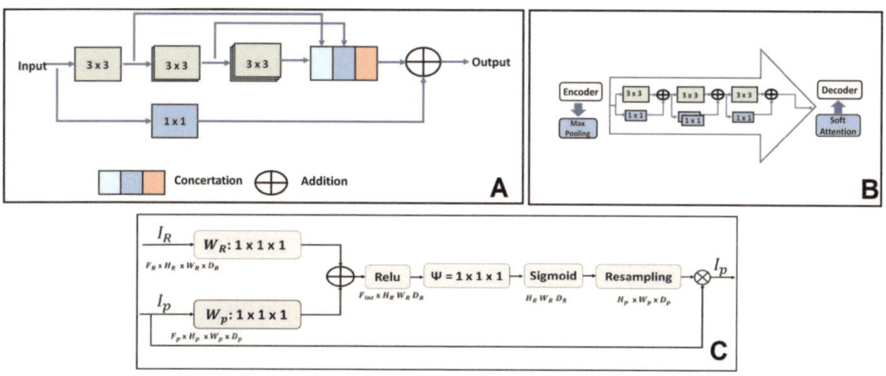


**Figure 4.**
*Description of (a) MultiRes block, (b) ResAtt-path and (c) attention gated block of MultiRes-attention U-Net architecture.*

The ResAtt path connects the U-Net encoder at each level to the attention modules in the decoding section of U-Net.

### 2.7 Self-attention

Soft attention gates (AGs) proposed by Oktay et al. [20] assist the model to focus on regions of interest by suppressing irrelevant location-based feature activations. AGs ensure that only salient spatial information is carried across skip connection which improves the network performance in false positives reduction. Soft attention gates (AGs), as shown in **Figure 3**(c), and illustrated in Eq. (1) contains two inputs (i) $I_p$—lower-level block input and, (ii) $I_R$—ResAtt-Path from the proposed skip connection layer. $I_p$ input is fed into 1×1 convolution filter for upsampling to match the dimensions of the inputs as illustrated in Eq. (2). The dimension matched inputs $x_{attention}$ *and* $x_{upsampled}$ are combined and passing through a ReLU activation function and sigmoid activation functions to yield a coefficients with values between 0 and 1.

Finally, these coefficients are upsampled through trilinear interpolation to generate the soft attention feature map. Which is then multiplied by the ResAtt-Path's skip connection to produce the final output as shown in Eq. (3)

$$x_{attention} = \text{Soft Attention}\left(I_p, I_R\right) \tag{1}$$

$$x_{upsampled} = Upsample\left(I_p\right) \tag{2}$$

$$output = ConvBlock\left(concat\left(x_{attention}, x_{upsampled}\right)\right) \tag{3}$$

### 2.8 Loss function

Segmentation model performance not only depends on the architecture of the network but also on the choice of the loss function [27] particularly in the scenario where there is a high-class imbalance. As we observed imbalance in SSAT, DSAT, and VAT distributions, we identified focal dice loss function as an appropriate loss function that handles class imbalance issues. The focal dice loss incorporates the focal loss where $\gamma = 0.5$ Eq. (4) and dice loss Eq. (5) together making it a robust loss function for the imbalanced class problems. It makes use of weighted components for each class based on their representation.

$$Focal\ loss = -(1 - \rho_t)^{\gamma} \log(\rho_t) \tag{4}$$

$$\text{Dice loss} = 1 - \text{dice coefficient} = 1 - \frac{2 * (A \cap B)}{A + B} \tag{5}$$

### 2.9 Post processing

Fat sub-region volumetric analysis & sub-region volume percentage is computing using Eqs. (6) and (7)

$$Vr = (TP_{ssat} + TP_{dsat} + TP_{vat}) * Ir * 1000 \tag{6}$$

where $TP_{ssat}, TP_{dsat}, TP_{vat}$ correspond to predicted voxel count of SSAT, DSAT and VAT classes & $Ir$ corresponds to each subject's voxel resolution. Sub-regions volumes percentage is computed using Eq. (7), where $TPi$ is the true positive volume of class $i$, and $\sum TPv$ is the total volume of the fat region.

$$\%Vc = \frac{TPi}{\sum TPv} * 100 \tag{7}$$

## 2.10 Training parameters

Single contrast fat-only 3D MR Dixon scans were converted to 2D slices for training (approximately 8000, 2D slices). Training was conducted on ubuntu 18.04 LTS operating system with NVIDIA Titan X GPU card with code written using TensorFlow framework [28] with hyperparameters of MultiRes-Attention U-Net is shown in **Table 1**.

## 2.11 Performance analysis

Multiclass Dice ratio (DR) & Hausdorff distance were two performance matrices used to evaluate the fat subregions segmentation which comprising of SSAT, DSAT and VAT regions.

The similarity between predicted and ground truth segmentation results is assessed by measuring the overlap using multiclass Dice score as illustrated in Eq. (8).

$$DSI_k = \frac{\sum(Ipred[Igt == k] == k) * 2.0}{\sum(Ipred[Ipred == k] == k) + \sum(Igt[Igt == k] == k)} \tag{8}$$

where $DSI_k$ is the subclass DSI value ranging between 0 and 1, where 1 means complete overlop of subregion, *Ipred* is the predicted output, *Igt* is the ground truth, and *k* is the number of classes.

Hausdorff Distance (HD) measures as the distance between two compact non-empty subsets of a metric space [30]. In order to find similarity between predicted (Pred) and ground truth (GT) HD measure between two closed and bounded subsets A and B of a given metric space M is defined as.

$$HD(Pred, GT) = \max(h(Pred, GT), h(GT, Pred)) \tag{9}$$

$$h(Pred, GT) = \max(dist(\alpha Pred, GT)) \tag{10}$$

$$dist(\alpha Pred, GT) = \min(\mu(\alpha Pred, GT)) \tag{11}$$

| Training parameters | Value |
|---|---|
| Number of filters at each levels of U-Net | 16,32,64,128,256 |
| Epochs | 150 |
| Optimizer | ADAM [29] |
| Learning rate | 0.00001 |
| Loss function | Focal dice loss |
| Weighted decay | 2e-6 |
| Dropout | 0.05 |
| Patience | 15 |

**Table 1**
*Illustrating the hyperparameters values in training MultiRes-attention U-Net.*

where $HD(Pred, GT)$ is the direct distance between Predicted region and ground truth, $dist(\alpha Pred, GT)$ is the distance from point to region GT and $\mu(\alpha, GT)$ is a point distance in the metric space. The smaller $\text{HD}(Pred, GT)$ indicates better segmentation accuracy i.e., less mismatch area.

## 3. Results

Accurate fat depot segmentation plays a significant role in evaluating fat distribution which can be used as biomarkers to assess metabolic syndrome and obesity. **Table 2** illustrates the training and testing Dice statistical index (DSI) (Mean $\pm$ SD) for MultiRes-Attention U-Net, MultiResUNet, and standard U-Net's 3-class (Class 1: Superficial Fat, Class 2: Deep-Superficial Fat, Class 3: Visceral fat) segmentation accuracies with trained on focal dice loss functions.

Dice score (**Table 1**) indicated that all the models show improved segmentation accuracy when trained under focal dice loss function.

## 4. Discussion

The removal of the arm region is an important step in pre-processing as it contains SAT, which may interfere with automatic segmentation. MR-based data augmentation techniques were used to increase the training samples and improve the generalization of the model. In this study, we have proposed a MultiRes-attention U-Net for the segmentation of the three abdominal fat compartments namely superficial subcutaneous fat, deep subcutaneous fat and visceral fat.. Algorithm took about 5 s to accurately segment and quantitate all the 3 different fat compartments thus reducing the time significantly. This enables the usage of our algorithm for clinical routines and large clinical trials.

Based on **Table 1**, the proposed algorithm performs better and provides a more accurate segmentation output than MultiResUNet due to the introduction of the AG module. Introduction of the attention module improved the identification of significant

| **DSI score for training (focal dice loss)** | **SSAT** | **DSAT** | **VAT** |
|---|---|---|---|
| U-Net | 0.9090±0.023 | 0.8727±0.035 | 0.8048±0.113 |
| MultiResUNet | 0.9751 ± 0.021 | 0.9732 ± 0.023 | 0.9679 ± 0.017 |
| MultiRes-Attention U-Net | **0.9877±0.022** | **0.9852±0.024** | **0.9758±0.022** |
| **DSI Score for Testing** | **SSAT** | **DSAT** | **VAT** |
| U-Net | 0.9071±0.020 | 0.8660±0.043 | 0.7426±0.140 |
| MultiResUNet | 0.9706±0.030 | 0.9657 ± 0.035 | 0.9586 ± 0.017 |
| MultiRes-Attention U-Net | **0.9781±0.029** | **0.9718±0.0349** | **0.9711±0.015** |
| **Hausdorff Dist** | **SSAT HD (mm)** | **DSAT HD (mm)** | **VAT HD (mm)** |
| U-NET | 4.8385 ± 0.023 | 4.5830 ± 0.4202 | 5.5176 ± 0.113 |
| MultiResUNet | 4.232 ± 0.121 | 4.323 ± 0.3242 | 4.332 ± 0.765 |
| MultiRes-Attention U-Net | 4.132 ± 0.868 | 4.199 ± 0.656 | 4.223 ± 0.133 |

**Table 2.**
*Performance comparison of models.*

features such as fascia boundary and smaller VAT components around the spine and preventing the network from learning false positive information. Focal dice loss function was found to be more appropriate in improving the overall segmentation results compared to cross-entropy (CE) loss and dice. Experimental results showed that focal-dice loss function could handle inherent class imbalance (amount of SSAT/DSAT/VAT in different slices) where cross-entropy or dice loss functions failed. The mean focal dice loss DSI for the test dataset was about 97.81% for SSAT, 97.18% for DSAT, and 97.11% for VAT, which is a significant improvement by 7%, 11%, and 23% respectively when compared to standard U-Net results. AHD of the proposed architecture is slightly better than MultiResUNet and when compared to standard U-Net, it is significantly better for 3 classes (SSAT, DSAT, and VAT). In addition, the model was able to separate SAT into SSAT and DSAT in lean subjects (broken or invisible fascia) and obese subjects (multiple fasciae). As shown in **Figure 5**, the model was also able to differentiate between VAT and bones, especially in the spine and pelvic regions. Further, MultiRes-Attention U-Net was tested on a new 190 data sets (unseen during training; upper & lower abdomen scans with different resolution) as illustrated in **Figure 6** which yielded accurate results for SSAT and DSAT but had few false positives in sacrum region VAT.

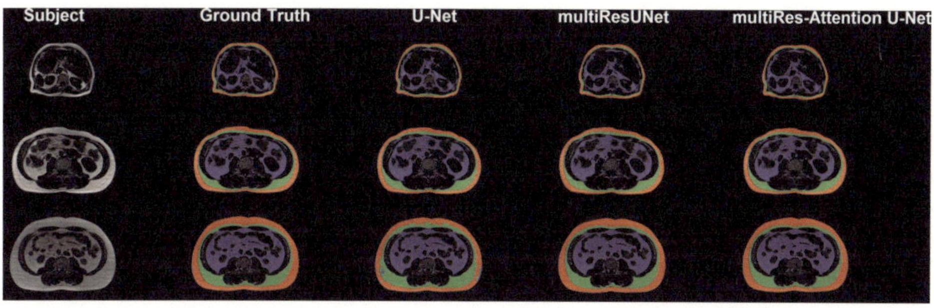


**Figure 5.**
*Shows comparison of predicted results of U-Net, MultiResUNet, and MultiRes-attention U-Net (loss function: Focal dice) on low-medium and high-fat subjects.*

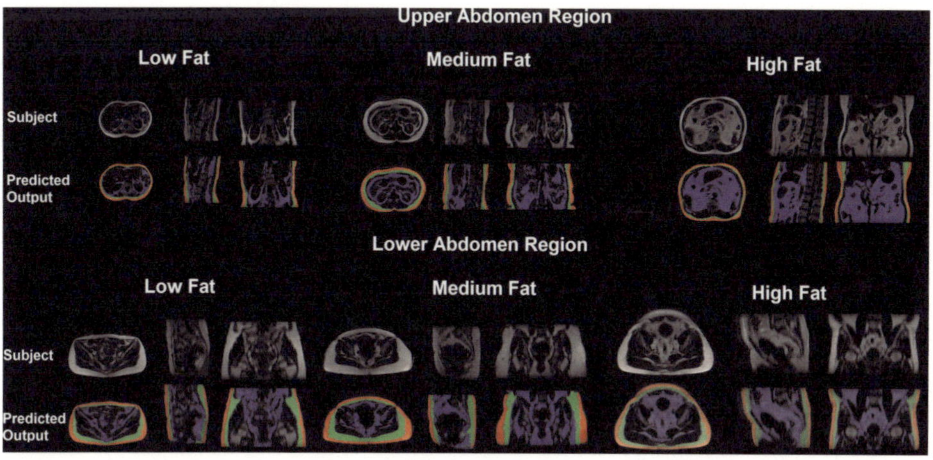


**Figure 6.**
*Illustration of the predicted result of MultiRes-attention U-Net on a few selected samples of new 190 data sets (unseen during training; upper & lower abdomen scans with different resolution).*

## 5. Conclusion

In this study, we propose MultiRes-Attention U-Net with hybrid loss function for segmentation of superficial and deep subcutaneous adipose tissue (SSAT & DSAT), and visceral adipose tissue (VAT) from abdominal MR scans. MultiRes block, ResAtt-Path, and attention gates can handle shape, scale, and heterogeneity in the abdominal data. Model performance is also dependent on the loss function, especially when there is data imbalance. In this research work, focal dice loss function compared to cross-entropy (CE) loss and dice were found to be more appropriate in improving the overall segmentation results. The proposed pipeline contains pre-processing, data augmentation, and automatic segmentation of fat compartments and fat quantification. The proposed algorithm takes less than 5 s for segmentation and quantification of 3 fat compartments are provided more generalizable results where the model was able to separate SAT into SSAT and DSAT in lean subjects (broken or invisible fascia) and in obese subjects (multiple fasciae) and also differentiate small VAT tissue from bones making it feasible for use in large clinical trials and clinical routine.

## Author details

Bhanu K.N. Prakash[1*], Arvind Channarayapatna Srinivasa[1], Ling Yun Yeow[1], Wen Xiang Chen[2], Audrey Jing Ping Yeo[3], Wee Shiong Lim[3] and Cher Heng Tan[2]

1 Bioinformatics Institute (BII), Agency of Science, Technology and Research (A*STAR), Singapore, Republic of Singapore

2 Department of Diagnostic Radiology, Tan Tock Seng Hospital, Singapore, Republic of Singapore

3 Department of Geriatric Medicine, Tan Tock Seng Hospital, Singapore, Republic of Singapore

*Address all correspondence to: bhanu_prakash@bii.a-star.edu.sg

## References

[1] Web page: who news: https://www.who.int/news-room/fact-sheets/detail/obesity-and-overweight.

[2] Tremmel M, Gerdtham UG, Nilsson PM, Saha S. Economic Burden of Obesity: A Systematic Literature Review. International Journal of Environmental Research and Public Health. 2017 Apr 19; **14**(4):435. DOI: 10.3390/ijerph14040435

[3] Brons C, Grunnet LG. Mechanisms in endocrinology: Skeletal muscle lipotoxicity in insulin resistance and type 2 diabetes: A causal mechanism or an innocent bystander? European Journal of Endocrinology. 2017;**176**:R67-R78. DOI: 10.1530/EJE-16-0488

[4] St-Pierre J, Lemieux I, Vohl MC, Perron P, Tremblay G, Despres JP, et al. Contribution of abdominal obesity and hypertriglyceridemia to impaired fasting glucose and coronary artery disease. The American Journal of Cardiology. 2002; **90**:15-18

[5] Chan JM, Rimm EB, Colditz GA, Stampfer MJ, Willett WC. Obesity, fat distribution, and weight gain as risk factors for clinical diabetes in men. Diabetes Care. 1994;**17**:961-969

[6] Seabolt LA, Welch EB, Silver HJ. Imaging methods for analyzing body composition in human obesity and cardiometabolic disease. Annals of the New York Academy of Sciences. 2015; **1353**:41-59. DOI: 10.1111/nyas. 12842

[7] Baum T, Cordes C, Dieckmeyer M, Ruschke S, Franz D, Hauner H, et al. MR-based assessment of body fat distribution and characteristics. European Journal of Radiology. 2016;**85**:1512-1518. DOI: 10.1016/j.ejrad.2016.02.013

[8] Schar M, Eggers H, Zwart NR, Chang Y, Bakhru A, Pipe JG. Dixon water-fat separation in PROPELLER MRI acquired with two interleaved echoes. Magnetic Resonance in Medicine. 2016;**75**:718-728. DOI: 10.1002/mrm.25656

[9] Positano V, Gastaldelli A, Sironi AM, Santarelli MF, Lombardi M, Landini L. An accurate and robust method for unsupervised assessment of abdominal fat by MRI. Journal of Magnetic Resonance Imaging. 2004;**20**:684-689. DOI: 10.1002/jmri.20167

[10] Demerath EW, Ritter KJ, Couch WA, Rogers NL, Moreno GM, Choh A, et al. Validity of a new automated software program for visceral adipose tissue estimation. International Journal of Obesity. 2007;**31**:285-291

[11] Kullberg J, Angelhed JE, Lonn L, Brandberg J, Ahlstrom H, Frimmel H, et al. Whole-body T1 mapping improves the definition of adipose tissue: Consequences for automated image analysis. Journal of Magnetic Resonance Imaging. 2006;**24**:394-401. DOI: 10.1002/jmri.20644

[12] Chew J, Yeo A, Yew S, Tan CN, Lim JP, Hafizah Ismail N, et al. Nutrition Mediates the Relationship between Osteosarcopenia and Frailty: A Pathway Analysis. Nutrients. 2020 Sep 27;**12**(10): 2957. DOI: 10.3390/nu12102957

[13] Kn BP, Gopalan V, Lee SS, Velan SS. Quantification of abdominal fat depots in rats and mice during obesity and weight loss interventions. PLoS One. 2014;**9**:e108979. DOI: 10.1371/journal.pone.0108979

[14] McBee MP, Awan OA, Colucci AT, Ghobadi CW, Kadom N, Kansagra AP, et al. Deep Learning in Radiology.

Academic Radiology. 2018 Nov;**25**(11): 1472-1480. DOI: 10.1016/j.acra.2018.02.018

[15] Grainger AT, Krishnaraj A, Quinones MH, Tustison NJ, Epstein S, Fuller D, et al. Deep learning-based quantification of abdominal subcutaneous and visceral fat volume on CT images. Academic Radiology. 2021; **28**(11):1481-1487. DOI: 10.1016/j.acra.2020.07.010 Epub 2020 Aug 6

[16] Nandakumar G, Srinivasan G, Kim H, Pi J. Comprehensive End-to-End Workflow for Visceral Adipose Tissue and Subcutaneous Adipose Tissue quantification: Use Case to improve MRI accessibility. In: 2020 IEEE 20th International Conference on Bioinformatics and Bioengineering (BIBE), Cincinnati, OH, USA, 2020. pp. 1060-1064. DOI: 10.1109/BIBE50027.2020.00179

[17] Estrada S, Lu R, Conjeti S, Orozco-Ruiz X, Panos-Willuhn J, Breteler MM, et al. FatSegNet: A fully automated deep learning pipeline for adipose tissue segmentation on abdominal Dixon MRI. Magnetic Resonance in Medicine. 2019; **83**:1471-1483

[18] Nowak S, Theis M, Wichtmann BD, Faron A, Froelich MF, Tollens F, et al. End-to-end automated body composition analyses with integrated quality control for opportunistic assessment of sarcopenia in CT. European Radiology. 2022 May;**32**(5): 3142-3151. DOI: 10.1007/s00330-021-08313-x

[19] Küstner T, Hepp T, Fischer M, Schwartz M, Fritsche A, Häring HU, et al. Fully Automated and Standardized Segmentation of Adipose Tissue Compartments via Deep Learning in 3D Whole-Body MRI of Epidemiologic Cohort Studies. Radiol Artif Intell. 2020 Oct 28;**2**(6):e200010. DOI: 10.1148/ryai.2020200010

[20] Oktay O, Schlemper J, Folgoc LL, Lee MJ, Heinrich MP, Misawa K, et al. Attention U-Net: Learning where to look for the pancreas. ArXiv abs/1804.03999. 2018

[21] Kafali SG, Shih SF, Li X, Chowdhury S, Loong S, Barnes S, et al. 3D Neural Networks for Visceral and Subcutaneous Adipose Tissue Segmentation using Volumetric Multi-Contrast MRI. Annual International Conference of the IEEE Engineering in Medicine & Biology Society (EMBC). 2021 Nov;**2021**:3933-3937. DOI: 10.1109/EMBC46164.2021.9630110

[22] Ibtehaz N, Rahman MS. MultiResUNet: Rethinking the U-Net architecture for multimodal biomedical image segmentation. Neural Networks: The Official Journal of the International Neural Network Society. 2020;**121**:74-87

[23] Bhanu PK, Arvind CS, Yeow LY, Chen WX, Lim WS, Tan CH. CAFT: a deep learning-based comprehensive abdominal fat analysis tool for large cohort studies. MAGMA. 2022 Apr;**35** (2):205-220. DOI: 10.1007/s10334-021-00946-9

[24] Ronneberger O, Fischer P, Brox T. U-Net: Convolutional networks for biomedical image segmentation. ArXiv 1505.04597. 2015

[25] He F, Liu T, Tao D. Why ResNet works? Residuals generalize. IEEE Transactions on Neural Networks and Learning Systems. 2020;**31**:5349-5362

[26] Cao Y, Liu S, Peng Y, Li J. DenseUNet: Densely connected UNet for electron microscopy image segmentation. IET Image Processing. 2020;**14**:2682-2689

[27] Sudre CH, Li W, Vercauteren T, Ourselin S, Cardoso MJ. Generalised dice overlap as a deep learning loss function for highly unbalanced segmentations. In: Deep Learning in Medical Image Analysis and Multimodal Learning for Clinical Decision Support. Springer; 2017. pp. 240-248

[28] Braiek HB, Khomh F. TFCheck : A TensorFlow Library for Braiek, Houssem Ben and Foutse Khomh. TFCheck : A TensorFlow Library for Detecting Training Issues in Neural Network Programs. In: 2019 IEEE 19th International Conference on Software Quality, Reliability and Security (QRS). 2019. pp. 426-433

[29] Kingma DP, Ba J. Adam: A method for stochastic optimization. ArXiv 1412.6980. 2015

[30] Andreev A, Kirov N. Hausdorff distances for searching in binary text images. Serdica Journal of Computing. 2009;**3**(1):23-46

Chapter 5

# Deep Learning for Natural Language Processing

*Yuan Wang, Zekun Li, Zhenyu Deng, Huiling Song and Jucheng Yang*


## Abstract

With the constantly growing number of topical or sentiment-bearing texts and dialogs on the Web, the demand for automatic language or text analysis algorithms continues to expand. This chapter discusses about advanced deep learning techniques for classical and hot research directions in the field of natural language processing, including text classification, sentiment analysis, and task-oriented dialog systems. In text classification, we focus on tasks of multi-label text classification and extreme multi-label text classification, which allow for automatically annotates the texts with the most relevant labels. In sentiment analysis, we look into aspect-based sentiment analysis that makes automatic extraction of fine-grained sentiment information from texts, and multimodal sentiment analysis that classifies people's opinions or attitudes from multimedia data through fusion techniques. In dialog system, we introduce how deep learning techniques work in pipeline mode and end-to-end mode for task-oriented dialog system. In this chapter, the rapidly evolving state of the research on the three topics is reviewed. Furthermore, trends in the research on deep learning for natural language processing are identified, and a discussion about future advances is provided.


**Keywords:** deep learning, text classification, sentiment analysis, task-oriented dialog system, tasks and models

## 1. Introduction

Deep learning becomes increasingly important due to the fast growing of internet contents and the urgent needs of big data in natural language processing (NLP).

The text classification task is one of the most fundamental scenarios in natural language processing (NLP), where the user enters the text and the model divides the input text into defined categories. Text classification tasks can be divided into multi-class text classification, multi-label text classification, hierarchical text classification and extreme multi-label text classification. In the multi-class text classification settings, there are two or more label categories in the label set, and each sample has only one relevant label. In the multi-label text classification (MLTC) settings, a sample may have one or more relevant labels. The hierarchical text classification is a special multi-class text task or multi-label task, where the labels have a hierarchical relationship between them. The extreme multi-label text classification task (XMTC) is annotating the most relevant labels for the text from a large label set with millions, or even billions, of labels. It is a limitation of traditional models that words are treated as independent features out of context. Deep learning methods have had great success in other related fields by automatically extracting context-sensitive features from raw text. Text classification techniques can be applied into problem classification [1], topic classification [2], and emotion classification [3]. Text classification tasks can be divided into the recommendation system domain, the legal domain, and the ad placement domain depending on the target domain. In the field of recommendation systems, predicting how much a user prefers a particular item. In the legal field, MLTC questions are used to predict the final outcome of bills. In the field of ad placement, personalized ads are tailored to users by inferring their characteristics and personal interests on social media.

Sentiment analysis refers to mining people's opinions and emotional attitudes toward various matters through modal information such as texts and images. In the early days, sentiment analysis was mainly used to analyze user reviews of products sold online, and thus confirm user preferences for purchasing products. With the popularity of self-publishing nowadays, sentiment analysis is more often used to identify the sentiment analysis of topic participants, to mine the value of topics, and to analyze related public opinion. Sentiment analysis has important application value for both society and individuals.

The dialog system relies on deep learning technology to act as an assistant to talk or chat with people to people. Task-oriented dialog system is used to solve specific problems in specific fields, such as movie ticket reservation, restaurant table reservation, etc. Because of its huge commercial value, it has attracted more and more people's attention.

This chapter is organized as follows: Section 2 discusses advancement in text classification, Section 3 outlines the sentiment analysis, Section 4 presents the task-oriented dialog system, and finally, Section 5 concludes the chapter.

## 2. Advancement in text classification

### 2.1 Multi-label text classification

There are three problems in MLTC settings. The process of obtaining comprehensive supervisory information is time-consuming and labor-intensive. The lack of theoretical support for the interpretability aspect of deep learning is also an issue that needs to be addressed. Modeling label dependencies is a major difficulty (**Figure 1**).

Multi-label text classification includes text pre-processing, text representation work using feature engineering, and classifier. Text pre-processing is a series of processes on the original text including word segmentation, cleaning, normalization, and so on. Text representation processes words into vectors or matrices so that computers can process them. Feature engineering is divided into heuristics, machine

**Figure 1.**
*Deep learning in multi-label text classification.*

learning-based methods, and deep learning-based methods. Deep learning-based approaches can be divided into text-based representations [4] and interactive representations [4] based on text and labels, depending on whether the model introduces labels information to represent the text.

### *2.1.1 Text representation*

Deep learning-based approaches can be divided into text-based representations [4] and interactive representations [4]. Text-based representations focus on converting text into machine-understandable form for subsequent natural language processing tasks. Interactive representations, on the other hand, focus on modeling dialog history and context to better understand the current dialog by considering different sentences in the dialog history and changes in user intent. It should be noted that text-based and interactive representations are not mutually exclusive but can be used in combination. In some tasks, text-based representations can be used first to convert individual texts into representation vectors, and then considered in conjunction with interactive representations to take into account contextual information for more accurate and comprehensive text comprehension and processing. For text-based representations, TextCNN [5] applies convolutional neural networks and uses multiple kernels of different sizes to extract key information in sentences. For interactive representations, LEAM [6] establishes the semantic interaction matrix between texts and labels to obtain the attention weight, so as to obtain the most relevant labels.

### *2.1.2 Deep learning models*

Deep learning-based text representation works to automatically acquire textual information, including word vector models and neural network models.

Word vector models based on distributed representations map vectors in high-dimensional space to low-dimensional space, alleviating the problem of feature sparsity. Commonly used word vectors include static word vectors word2vec [7], global vectors for word representation (Glove) [8], dynamic word vector models such as embedding from language models (ELMo) [9], and bidirectional encoder representations from transformers (BERT) [10] models. Word2vec can further subdivided into CBOW [7] and skip-gram. The input to the CBOW [7] is a vector of neighboring words of a central word, and the output is a vector of words of that central word. The input to the skip-gram model is a vector of central words, and the output is a vector representation of the surrounding words of that central word. This is generally better than CBOW. Glove [8] statistical co-occurrence matrix and sliding window, taking into account both local and global information. Firstly, the co-occurrence matrix is constructed by using the corpus, and secondly, the relationship between the word vector and the co-occurrence matrix is constructed. ELMo [9] has a three-layer structure, with the first layer being the word2vec or Glove, and the next two layers being the two bidirectional long- and short-term memory (Bi-LSTM) extracting word contextual features to effectively solve the problem of multiple meanings of words. BERT uses transformer as the main framework for capturing bidirectional relations in utterances and constructs mask language model and next sentence prediction as targets for multi-task training in terms of training tasks.

Common neural network models include convolutional neural networks (CNN) [11], recurrent neural network (RNN) [12], long- and short- term memory network (LSTM) [1], and attention mechanisms [13]. CNN sets different convolutional kernels to extract local contextual information of the text and deepens the multi-layer convolutional and pooling layers to capture deeper textual information. In detail, the input layer obtains low-dimensional word vectors. The convolution layer extracts the local information of the text and the pooling layer reduces the feature dimension and prevents overfitting. Finally, the text and label dimensions are unified by the fully connected layer. The softmax layer is normalized to obtain the probability. RNN uses time series memory history information to obtain a representation of text content information by accepting text sequences of arbitrary length and generating a fixed-length vector. Gradient vanishing or explosion prevents RNN from effectively learning long-term dependencies and correlations. LSTM, in order to solve the problem of RNN on long-term dependency, adds forgetting gates, input gates, and output gates units to RNN to avoid gradient vanishing or explosion. The methods above assign the same weight to words and cannot distinguish the importance of words. Inspired by human attention, the attention mechanism is introduced to focus on key information and key contents, making it easy for models to focus on the weighted part and improve the classification accuracy. The attention mechanisms are usually divided into three categories, namely local attention, global attention, and self-attention mechanisms. Global attention considers entire text of words, assigning weights between 0 and 1 to obtain the text representation. Local attention assigns a weight of either 0 or 1 to each word, discarding some irrelevant items directly. Self-attention assigns weights based on the interaction of input words, which has advantage of parallel computing in long text classification.

In conclusion, both word vector models and neural network models are important components of deep learning-based text representation techniques, and they each have their own advantages and can be selected according to the needs of specific tasks. Word vector models focus more on the static representation of words, while neural network models are better able to capture the dynamic information of the context. Word vector models are relatively fast to train, while neural network models usually require larger computational resources and longer training time. Neural network models may perform better on some complex tasks, but for some simple tasks, word vector models are effective enough.

## 2.2 Extreme multi-label text classification

Extreme multi-label text classification learns a classifier that labels the most relevant subset of labels for a document from a very large set of labels. The main challenge is the millions of labels, features, and training points. The current research architectures in extreme multi-label text classification can be divided into four main categories, namely one-vs-all models, embedding-based models, tree-based models, and deep learning models. Due to the high computational costs brought by large-scale labels, the existing MLTC techniques have difficulty solving the XMTC problem. It can be seen that the extreme label text classification task is trapped in a large label space and feature space, leading to two pressing problems. The first problem is the power-law distribution, where long-tailed labels have very little data associated with them, making it difficult to obtain dependencies between labels, presenting data sparsity and scalability in extreme text classification work. The second problem is that computation is expensive, and the same results can be obtained at less cost using data augmentation techniques. One-vs-all models train a separate classifier for each label on the entire datasets. The one-vs-all models usually classifies well and with high accuracy; however, it assumes that the individual labels are independent of each other and uncorrelated, resulting in a cost that grows linearly with the number of labels. Embedded models typically use the relationships between labels to map labels from a high-dimensional space to a low-dimensional space using a linear matrix mapping approach as a way to reduce the total number of parameters in the model and reduce the training time required for the model. The limitation of the embedding method is that it ignores the correlation between input and output, resulting in an unaligned embedding of the two. Tree-structured models are trained to produce instance or labeled trees to make predictions, such as decision trees, random forests, Hoffman trees, etc. Traditional tree-based approaches can harm performance due to large tree height and large cluster size.

All three types of models mentioned above are based on bag-of-words representations of text, where words are treated as independent features out of context and cannot capture deep semantic information. In contrast, deep learning models can automatically extract implicit contextual features from raw text for extreme multi-label text classification.

Typical work, such as XML-CNN [14], first explored the application of deep learning to XMTC, proposing a series of CNN models for XMTC, modeling convolutional neural networks and dynamic maximum pooling layers to extract semantic features of text, and introducing hidden bottleneck layers to reduce model parameters and accelerate training; however, XML-CNN [14] cannot capture the most important subtext of each label. Therefore, AttentionXML [15] solves this problem with two techniques. Firstly, a multi-label attention mechanism is introduced to capture the most relevant parts of text for each label. Secondly, a shallow and wide probabilistic label tree is built to handle millions of labels. Lightxml [16] adopts BERT as an encoder for text and obtains a better text representation, which is the state-of-the-art extreme multi-label text classification model. DeepXML [17] designed a framework to decompose XMTC into four subtasks using this framework. These four subtasks are optimized by selecting different components to generate a series of algorithms, including Astec [17], DECAF [18], GalaXC [19], and ECLARE [20]. Astec [17] needs to use label clustering to obtain intermediate feature representations. DECAF [18] jointly learn model parameters and feature representation to get label metadata. GalaXC [19] introduces a label attention mechanism to make more accurate predictions based on the multi-resolution embedding of nodes given by the graph. ECLARE [20] allows collaborative learning using label-label correlations.

In summary, one-vs-all models are simple and intuitive and can be used flexibly with a variety of binary classification algorithms but ignore the correlation between labels, which may lead to inaccurate classification. Embedding-based models capture semantic information but do not directly model the correlation between labels. Tree-based models are able to handle high-dimensional and nonlinear data and can capture correlations between nested features and labels. Deep learning models are capable of learning complex feature representations and contextual correlations and are suitable for large-scale data and complex tasks.

## 3. Advancement in sentiment analysis

This section will introduce the aspect-based sentiment analysis (ABSA) and multimodal sentiment analysis in the sentiment analysis task, which is a classical task in the field of natural language processing, and we will mainly introduce the deep learning techniques for sentiment analysis since they have better performance than the past machine learning methods and are the mainstream methods in the field of sentiment analysis.

### 3.1 Aspect-based sentiment analysis

The concept of ABSA was first introduced in 2010 by Thet et al. [21], and further, Liu [22] gave a definition of viewpoint in 2012; sentiment analysis and opinion mining refers to the field of research that analyzes people's opinions, sentiments, evaluations, attitudes, and emotions from written language. From 2014 to 2016, SemEval, an international semantic evaluation conference, has included the ABSA task as one of its subtasks and provided a series of benchmark datasets [23, 24], which have all been manually annotated. In recent years, the aspect-based sentiment analysis task has been receiving attention from many scholars, especially after the rapid application of deep learning and other related technologies in the fields of data mining, information retrieval, and intelligent question and answer. Therefore, research related to aspect-based sentiment analysis based on deep learning has also continued to achieve breakthroughs [25–29], and the ABSA task has gradually become one of the popular research topics in the field of NLP (**Figure 2**).

The advantage of aspect-based sentiment analysis is mainly that text sentiment analysis is fine-grained. Coarse-grained sentiment analysis can often only capture one-sided single sentiment tendency and cannot analyze detail from each attribute level. A review text often contains sentiment views for different evaluation objects, for example, "the service of this restaurant is good, but the taste is bad." The text of this review evaluates the two aspects of "service" and "taste" separately, and the document-level and sentence-level sentiment analysis cannot mine each aspect separately. Therefore, aspect-based sentiment analysis is needed for re view texts that contain multiple aspects [30, 31].

Sentiment analysis methods based on deep learning can be divided into fourmain types: sentiment analysis methods with a single neural network, sentiment analysis methods with a hybrid neural network, sentiment analysis with the introduction of attention mechanisms, and sentiment analysis using pre-trained models.

The main methods for sentiment analysis of single neural networks are introducing a series of neural network models [32, 33] (e.g., CNN, RNN, etc.). CNN is mainly used to extract local features of text data, abstract low-dimensional vectors into vector

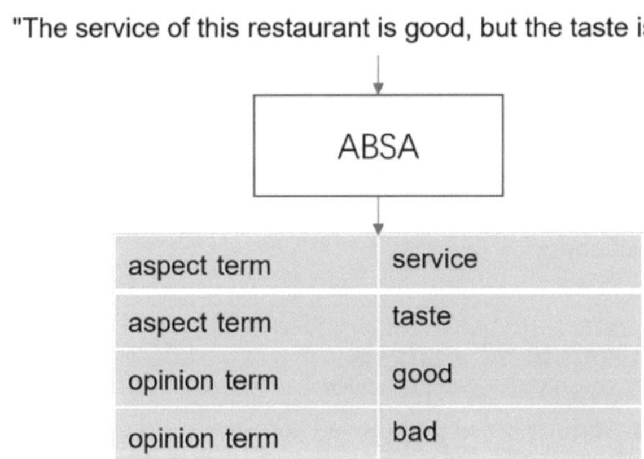


**Figure 2.**
*The working effect of ABSA.*

representations with high-level semantics after operations such as convolutional pooling, and then process the coded representations and output the results. Lu et al. [34] made full use of syntactic relations and sentiment dependency information and proposed an aspect-gated graph convolutional network (AGGCN) to implement aspect-based sentiment analysis work. Liang et al. [35] made full use of the dependency syntactic knowledge and designed a dependency-embedded graph convolutional network applied to end-to-end sentiment analysis. Wang et al. [36] proposed a new unified location-aware convolutional neural network (UP-CNN) to solve the problem of difficult to fully utilize aspect location information.

In ABSA tasks, attention mechanisms have received a lot of attention and have been actively used in aspect-based sentiment analysis tasks because of the different importance of information in different parts of the text for aspect-based sentiment analysis tasks, and attention mechanisms have ability to adaptively identify key information and enhance attention to it [37–40]. Liao et al. [41] use a two-way transformer-based RoBERTa model to extract features from text and aspect word strings and use a cross-attention mechanism to add attention to the most relevant features for a given aspect category.

### 3.2 Multimodal sentiment analysis

With the rapid development of information and network technology and the widespread use of mobile terminals, people are gradually showing a trend of diversifying the content they publish. The messages they publish for different events and topics are no longer limited to a single text form, but tend to publish multimodal content combining text and images to express their feelings and opinion aspect-based. This situation and trend have attracted academic attention to multimodal sentiment analysis research, and by analyzing the sentiment tendency implied by these multimodal data, it has great application value in box office prediction, product marketing, political election, product recommendation, mental health analysis, etc. Therefore, multimodal sentiment analysis has become a hot research topic in recent years [42, 43]. Multimodal sentiment analysis is the process of combining documents that describe the same thing in different forms (e.g., sound, image, text, etc.) to enrich our perception of the thing and analyze the sentiment it expresses. The term modality is generally associated in academic research with the sensory modalities that represent our primary communication and sensory channels, and when a research question or data set contains multiple modalities, it is characterized as a multimodal task or multimodal data set. In general, academics have focused on (but not limited to) three modalities: (1) natural language, both spoken and textual, (2) visual signals, often represented by images or videos, and (3) acoustic signals, such as intonation and audio. Multimodal learning is a dynamic multidisciplinary field that is breaking new ground in many tasks such as multimodal sentiment analysis, cross-modal retrieve, image caption, audiovisual speech recognition, and visual question and answer, visual speech recognition, and other tasks (**Figure 3**).

Multimodal sentiment analysis makes full use of data from different modalities for accurate sentiment prediction. In 2016, a cross-modality consistent regression (CCR) model was proposed in the literature [44]. The authors of this paper concluded that the overall sentiment of text and image unimodal, as well as multimodal is the same with respect to representation of modality, text including descriptions and captions of images, and learning visual features using CNNs, which outperformed the unimodal model. In the same year, work [45] proposed a tree-structured recursive neural

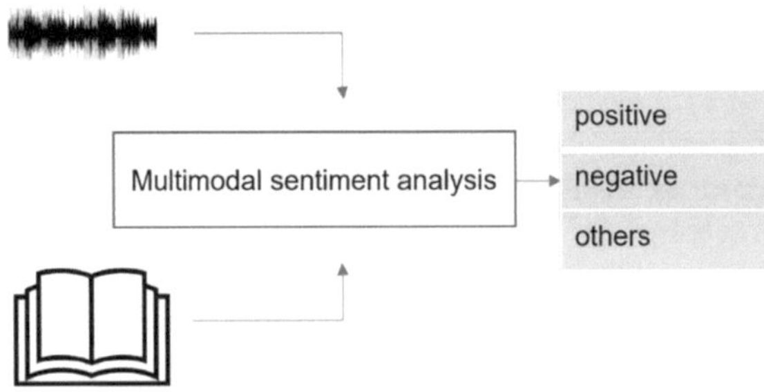


**Figure 3.**
*The working effect of MSA.*

networks (TreeLSTM) that use a tree structure and incorporates visual attention mechanisms. The system builds a structured structure based on sentence parsing aimed at aligning text words and image regions for accurate analysis and incorporates LSTM and attention mechanisms to learn a robust joint visual text representation with contemporaneous optimal results. In addition, the problem of image text mismatch and defects in social media data such as spoken words, misspellings, and lack of punctuation, pose a challenge to the task of sentiment analysis of multimodal data, and to address this challenge, in 2017, Xu et al. constructed different multimodal sentiment analysis networks, such as the hierarchical semantic attentional network (HSAN) [46] and multimodal deep semantic network (MultiSentiNet) [47]. HSAN focused on image captions and proposed a hierarchical semantic network model based on image captions in a multimodal sentiment analysis task using image captions to extract visual semantic features as additional information for text. MultiSentiNet, on the other hand, extracts image features from both objects and scenes and proposes a visual feature-guided attentional long- and short-term memory network to extract words that contribute to the understanding of text sentiment and aggregates these words with visual semantic features, objects and scenes. In 2018, co-memory network [48] proposed a novel co-memory network (CoMN), which models the interdependence between vision and text through memory networks to fully consider the interrelationship between multimodal data. In 2020, multi-view attentional network (MVAN) [49] utilizes a continuously updated memory network to obtain deep semantic features of images and texts. The authors found that existing datasets for multimodal sentiment analysis generally labeled only positive, negative and neutral sentiment polarities, and lacked graphical multimodal datasets for more detailed sentiment classification, so the authors constructed a large-scale image text multimodal dataset (TumEmo) based on social media multimodal data. Cheema proposed a simple and effective multimodal neural network (Sentiment Multi-Layer Neural Network, Se-MLNN) [50] model that used RoBERT to extract text features containing contextual features and multiple high-level image features from multiple perspectives to accurately predict the overall sentiment after fusing the features.

## 4. Advancement in task-oriented dialog system

This chapter introduces the task-oriented dialog system, including pipeline mode and end-to-end mode (**Figure 4**).

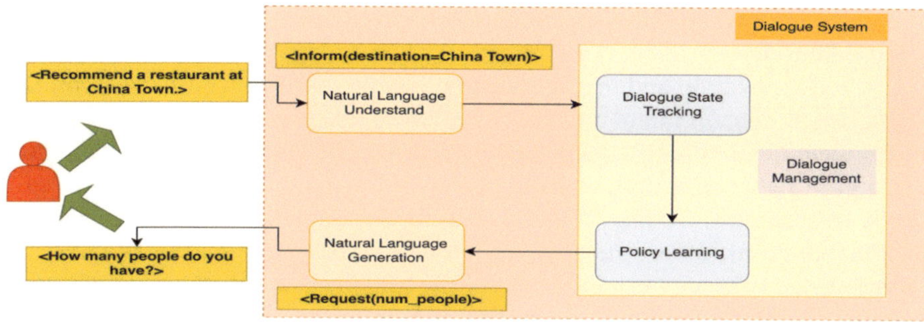


**Figure 4.**
*Task-oriented dialog system.*

## 4.1 Pipeline mode

Task-oriented dialog system aims to process user messages accurately and puts forward fairly requirements for response constraints. Therefore, a pipeline method is proposed to generate responses in a controllable way. It is mainly divided into four parts: natural language understanding, dialog state tracking, dialog strategy learning, and natural language generation. The natural language understanding module converts the original user messages into semantic slots and classifies the domain and user intentions. Dialog status tracking module iteratively calibrates the dialog status based on the current input and dialog history. The dialog state includes relevant user actions and slot value pairs. The dialog strategy learning module tracks the calibrated dialog state according to the dialog state and decides the next action of the dialog agent. Finally, the natural language generation module converts the selected conversation actions into natural language for feedback to users. For example, in the movie ticket reservation task, the agent interacts with the movie knowledge base to retrieve movie information with specific constraints [51], such as movie name, time, cinema, etc.

### *4.1.1 Natural language understanding*

Natural language understanding has a significant impact on the response quality of the whole system, which converts the user generated natural language messages into semantic slots and classified them. There are three tasks involved: domain classification, intention detection and slot filling. Domain classification aims to determine to which particular domain or topic the user input belongs. It categorizes the user's text into predefined domains, such as hotel booking, flight enquiry, weather information, etc. By identifying the subject domain to which the input relates, it can be passed to the appropriate processing module for further parsing. Intention detection refers to determining the user's intent or purpose in a particular domain. It focuses on the purpose behind the user's input rather than just the input text itself. For example, in the domain of hotel booking, a user may have different intentions, such as finding a hotel, booking a hotel, canceling a booking, etc. The goal of intent recognition is to identify the specific intent of the user so that the system can take the appropriate action or provide the correct response. Slot filling is the process of identifying and extracting key information from user input that is relevant to a specific domain. Slots are usually parameters or variables related to the intent, such as date, location, person's name, price, etc. Through slot filling, the system can capture and record the specific information provided by the user in a particular domain. For example, in a hotel reservation domain, slots may include check-in date, check-out date, location, room type, etc.

Domain classification and intent detection belong to the same classification task. The problem of domain intent and classification of dialog is solved through deep learning, including building a deep convex network [52], which combines the prediction of a prior network with the current dialog as the overall input of the current network. In order to solve the difficulty of using depth neural networks to predict fields and intentions, some scholars used restricted Boltzmann machines and depth belief networks to derive the parameters of the initialized depth neural networks [53]. In order to take advantage of the advantages of recurrent neural networks (RNN) in sequence processing, some work used recurrent neural networks as dialog encoders and predicted intentions and domain categories [54]. Some scholars have proposed a short text intention classification model. Due to the lack of information in a single conversation turn, it is difficult to identify the intention of phrases. Using RNN or CNN structure to fuse the dialog history, and obtain the context information as the additional input of the current turn information [55]. This model has achieved good performance in intention classification tasks. Recently, by pre-trained task-oriented dialog BERT, this method has achieved high accuracy in intention detection tasks. The proposed method can effectively alleviate the problem of data shortage in specific areas.

Slot filling, also known as semantic tagging problem, is a sequence classification problem. This model needs to predict multiple targets at the same time. Deep belief network shows good ability in deep structure learning. Some scholars built a sequence marker based on deep belief network. In addition to the named entity recognition input features used in traditional markers, they also combined part of speech and syntactic features as part of the input. Recurrent structures are beneficial to sequence marking tasks because they can track information along past time steps to maximize the use of sequence information. Some scholars first proposed that RNN language models can be applied to sequence tagging rather than simply predicting words [56]. At the output end of RNN, the sequence labels corresponding to the input words are not normal words. Some scholars further studied the impact of different recurrent structures on slot filling tasks and found that all RNN models are superior to the simple conditional random field method [57]. Because the shallow output representation of traditional semantic annotation lacks the ability to express structured dialog information, the slot filling task is regarded as a template based tree decoding process by iteratively generating and filling templates [58].

### *4.1.2 Dialog status tracking*

Dialog state tracking (DST) is the first module of the dialog manager. According to the entire dialog history, each turn tracks the user's goals and relevant details, providing the strategy learning module with the information needed for decision-making. There is a close relationship between natural language understanding and dialog state tracking. Both of them need to fill slots of dialog information [59]. However, they actually play two different roles. The natural language understanding module attempts to classify current user messages, such as intention recognition and domain recognition, and slots to which each message character belongs.

The first flow can be considered as a multi-class classification task. For multi-class classification DST, the tracker predicts to select the correct class from multiple values. Some scholars used RNN as a neural tracker to obtain the perception of dialog context

[60]. The tracker finally makes a binary prediction of the current slot value pair based on the dialog history. The second flow of neural tracker with unfixed slot names and values attracts more attention because it not only reduces the model and time complexity of DST tasks but also helps to train task-oriented dialog systems end-to-end. Some scholars proposed the belief span, that is. the text corresponding to the dialog context spans to a specific slot [61]. They built a two-stage CopyNet to copy and store the slot value history storage slot in the dialog to prepare for neural response. The belief span promotes the end-to-end training of the dialog system and improves the tracking accuracy outside the vocabulary. Based on this, some scholars proposed the minimum belief span, which is not scalable to generate belief state domains from scratch when the system interacts with APIs from different sources [62]. Some scholars proposed a trade model. The model also applies the replication mechanism and uses a soft-gated pointer generator to generate the slot value dialog context based on the domain slot pair and coding [63].

#### *4.1.3 Natural language generation*

Natural language generation is the last module in the pipeline mode of task-oriented dialog system. It tries to convert the dialog actions generated by the dialog manager into the final natural language representation. The standard flow of the defined natural language generation module is composed of four components, and its core components are content determination, sentence planning, and surface implementation.

The deep learning method is applied to further enhance the NLG performance, and the pipeline is folded into a single module. The generation of end-to-end natural languages has made gratifying progress and is the most popular way to implement NLG. Some scholars believed that natural language generation should be completely data-driven and not rely on any expert rules [64]. They proposed a statistical language model based on RNN, which uses semantic constraints and syntax trees to learn response generation. In addition, they also used CNN re-ranked to further select better answers. Similarly, some scholars used LSTM model to learn sentence planning and surface implementation at the same time. Some scholars used GRU to further improve the generation quality on multiple domains [65]. The proposed generator always generates high-quality responses on multiple domains. To improve the adaptability of the domain recurrent model, some scholars proposed to first train the recurrent language to model the data synthesized from the data sets outside the domain, and then fine-tune the relatively small data sets within the domain. This training strategy has proved to be effective in human assessment [66].

### 4.2 End-to-end mode

In the process of building an end-to-end task-oriented dialog system, a complex neural network model is used to implicitly represent key functions, and all modules are integrated into one module. The research of task-oriented end-to-end neural network model mainly focuses on training methods or model architecture, which is the key to response correctness and quality [67]. An incremental learning framework is proposed to train their end-to-end task-oriented system. The main idea is to establish an uncertainty estimation module to evaluate the confidence of the generated response. If the confidence is higher than the threshold value, the response will be accepted. If the confidence score is lower, the manual response will be introduced.

Recent works often do not build end-to-end systems to apply in a pipeline manner. Instead, they use complex neural models to implicitly represent key functions and integrate modules into one. Task-oriented end-to-end neural model research focuses on training methods or model architecture, which is the key and quality of response correctness. Some scholars proposed an incremental learning framework to train their end-to-end learning task-oriented system [61]. The main idea is to establish an uncertainty evaluation module to evaluate the confidence of the generated appropriate response. If the confidence score is higher than the threshold, then the response will be accepted, while if the confidence score is very low. The agent can also use online learning to learn from human responses. Some scholars use model agnostic meta learning (MAML) to jointly improve adaptability and reliability [68]. In real life online service tasks, there are only a few training samples. Similarly, some scholars also used MAML to train the end-to-end neural model to promote domain adaptation, which enables the model to train rich resource tasks first, and then train limited new task data [59]. Other scholars trained an inconsistent order detection module in an unsupervised manner [63]. The module detects whether the command discourse generates a more coherent response.

## 5. Conclusions

Most existing shallow and deep learning models have structures that can be used for text classification, including integrated approaches. BERT learns a form of linguistic representation that can be used to fine-tune many downstream NLP tasks. The main approaches are to add data, increase computational power, and design training programs to obtain better results. The trade-off between data and computational resources and predictive performance is worth investigating. Due to the inability to collect data with full supervisory information, so MLTC is gradually turning to the problem of classification with limited supervised information. Since the excellent performance of AlexNet in 2012, deep learning has shown great potential. How to leverage the powerful learning capabilities of deep learning to better capture the label dependencies is key to solving MLTC tasks.

With the development of deep learning technology in the application of emotion analysis tasks, the performance of emotion analysis has been greatly improved. However, some tasks and scenarios still need more abundant data sets to evaluate the model more accurately.

Although deep learning has achieved remarkable results in the dialog system, in the pipeline mode, if accurate and fast access to user intentions is still the demand of the industry, in the end-to-end mode, controllability, and interpretability also need to be further studied.

*DOI: http://dx.doi.org/10.5772/intechopen.112550*

## Author details

Yuan Wang*, Zekun Li, Zhenyu Deng, Huiling Song and Jucheng Yang
College of Artificial Intelligence, Tianjin University of Science and Technology, China

*Address all correspondence to: wangyuan23@tust.edu.cn

## References

[1] Graves A. Long short-term memory. In: Supervised sequence labelling with recurrent neural networks. Berlin: Springer; 2012. pp. 37-45

[2] Sakai Y, Matsuoka Y, Goto M. Purchasing behavior analysis model that considers the relationship between topic hierarchy and item categories. In: International Conference on Human-Computer Interaction. Cham: Springer; 2022. pp. 344-358

[3] Chen Z, Qian T. Transfer capsule network for aspect level sentiment classification. In: Proceedings of the 57th Annual Meeting of the Association for Computational Linguistics. Washington: ACL; 2019. pp. 547-556

[4] Li Q, Peng H, Li J, Xia C, Yang R, Sun L, et al. A survey on text classification: From traditional to deep learning. ACM Transactions on Intelligent Systems and Technology (TIST). 2022;**13**(2):1-41

[5] Chen Y. Convolutional neural network for sentence classification. [Master's thesis], University of Waterloo. 2015

[6] Wang G, Li C, Wang W, Zhang Y, Shen D, Zhang X et al. Joint embedding of words and labels for text classification. arXiv preprint arXiv: 1805.04174. 2018

[7] Mikolov T, Chen K, Corrado G, Dean J. Efficient estimation of word representations in vector space. arXiv preprint arXiv:1301.3781. 2013

[8] Pennington J, Socher R, Manning CD. Glove: Global vectors for word representation. In: Proceedings of the 2014 Conference on Empirical Methods in Natural Language Processing (EMNLP). Toronto: ACL; 2014. pp. 1532-1543

[9] Sarzynska-Wawer J, Wawer A, Pawlak A, Szymanowska J, Stefaniak I, Jarkiewicz M, et al. Detecting formal thought disorder by deep contextualized word representations. Psychiatry Research. 2021;**304**:114135

[10] Devlin J, Chang M-W, Lee K, Toutanova K. Bert: Pre-training of deep bidirectional transformers for language understanding. 2018. arXiv preprint arXiv:1810.04805

[11] Kalchbrenner N, Grefenstette E, Blunsom P. A convolutional neural network for modelling sentences. arXiv preprint arXiv:1404.2188. 2014

[12] Zaremba W, Sutskever I, Vinyals O. Recurrent neural network regularization. arXiv preprint arXiv:1409.2329. 2014

[13] Vaswani A, Shazeer N, Parmar N, Uszkoreit J, Jones L, Gomez AN, et al. Attention is all you need. Advances in Neural Information Processing Systems. 2017;**30**

[14] Liu J, Chang W-C, Wu Y, Yang Y. Deep learning for extreme multi-label text classification. In: Proceedings of the 40th International ACM SIGIR Conference on Research and Development in Information Retrieval. New York: ACM; 2017. pp. 115-124

[15] You R, Zhang Z, Wang Z, Dai S, Mamitsuka H, Zhu S. Attentionxml: Label tree-based attention-aware deep model for high-performance extreme multi-label text classification. Advances in Neural Information Processing Systems. 2019;**32**

[16] Jiang T, Wang D, Sun L, Yang H, Zhao Z, Zhuang F. Lightxml:

Transformer with dynamic negative sampling for high-performance extreme multi-label text classification. In: Proceedings of the AAAI Conference on Artificial Intelligence. Vol. 35. Toronto: AAAI; 2021. pp. 7987-7994

[17] Dahiya K, Saini D, Mittal A, Shaw A, Dave K, Soni A, et al. Deepxml: A deep extreme multi-label learning framework applied to short text documents. In: Proceedings of the 14th ACM International Conference on Web Search and Data Mining. New York: ACM; 2021. pp. 31-39

[18] Mittal A, Dahiya K, Agrawal S, Saini D, Agarwal S, Kar P, et al. Decaf: Deep extreme classification with label features. In: Proceedings of the 14th ACM International Conference on Web Search and Data Mining. New York: ACM; 2021. pp. 49-57

[19] Saini D, Jain AK, Dave K, Jiao J, Singh A, Zhang R, et al. Galaxc: Graph neural networks with labelwise attention for extreme classification. In: Proceedings of the Web Conference 2021. New York: ACM; 2021. pp. 3733-3744

[20] Mittal A, Sachdeva N, Agrawal S, Agarwal S, Kar P, Varma M. Eclare: Extreme classification with label graph correlations. In: Proceedings of the Web Conference 2021. New York: ACM; 2021. pp. 3721-3732

[21] Thet TT, Na J-C, Khoo CSG. Aspect-based sentiment analysis of movie reviews on discussion boards. Journal of Information Science. 2010;**36**(6): 823-848

[22] Liu B, Zhang L. A survey of opinion mining and sentiment analysis. In: Aggarwal, C., Zhai, C. (eds) Mining Text Data. Boston, MA: Springer; 2012

[23] Pontiki M, Galanis D, Papageorgiou H, Manandhar S, Androutsopoulos I. Semeval-2015 task 12: Aspect based sentiment analysis. In: Proceedings of the 9th International Workshop on Semantic Evaluation (SemEval 2015). Toronto: ACL; 2015. pp. 486-495

[24] Pontiki M, Galanis D, Papageorgiou H, Androutsopoulos I, Manandhar S, Al-Smadi M, et al. Semeval-2016 task 5: Aspect based sentiment analysis. In: International Workshop on Semantic Evaluation. Toronto: ACL; 2016. pp. 19-30

[25] Do HH, Prasad PWC, Maag A, Alsadoon A. Deep learning for aspect-based sentiment analysis: A comparative review. Expert Systems with Applications. 2019;**118**: 272-299

[26] Akhtar MS, Gupta D, Ekbal A, Bhattacharyya P. Feature selection and ensemble construction: A two-step method for aspect based sentiment analysis. Knowledge-Based Systems. 2017;**125**:116-135

[27] Peng H, Ma Y, Li Y, Cambria E. Learning multi-grained aspect target sequence for chinese sentiment analysis. Knowledge-Based Systems. 2018;**148**: 167-176

[28] Tang F, Luoyi F, Yao B, Wenchao X. Aspect based fine-grained sentiment analysis for online reviews. Information Sciences. 2019;**488**:190-204

[29] Liu N, Shen B. Rememnn: A novel memory neural network for powerful interaction in aspect-based sentiment analysis. Neurocomputing. 2020;**395**: 66-77

[30] Xiao D, Ren F, Pang X, Cai M, Wang Q, He M, et al. A hierarchical

and parallel framework for end-to-end aspect-based sentiment analysis. Neurocomputing. 2021;**465**:549-560

[31] Zhou J, Zhao J, Huang JX, Qinmin Vivian H, He L. Masad: A large-scale dataset for multimodal aspect-based sentiment analysis. Neurocomputing. 2021;**455**:47-58

[32] Khasanah IN. Sentiment classification using fasttext embedding and deep learning model. Procedia Computer Science. 2021;**189**: 343-350

[33] Basiri ME, Nemati S, Abdar M, Cambria E, Rajendra U, Acharya. Abcdm: An attention-based bidirectional CNN-RNN deep model for sentiment analysis. Future Generation Computer Systems. 2021;**115**:279-294

[34] Qiang L, Zhu Z, Zhang G, Kang S, Liu P. Aspect-gated graph convolutional networks for aspect-based sentiment analysis. Applied Intelligence. 2021; **51**(7):4408-4419

[35] Liang Y, Meng F, Zhang J, Chen Y, Jinan X, Zhou J. A dependency syntactic knowledge augmented interactive architecture for end-to-end aspect-based sentiment analysis. Neurocomputing. 2021;**454**:291-302

[36] Wang X, Li F, Zhang Z, Guangluan X, Zhang J, Sun X. A unified position-aware convolutional neural network for aspect based sentiment analysis. Neurocomputing. 2021;**450**: 91-103

[37] Li Z, Li L, Zhou A, Hongbin L. Jtsg: A joint term-sentiment generator for aspect-based sentiment analysis. Neurocomputing. 2021;**459**:1-9

[38] Qiannan X, Zhu L, Dai T, Yan C. Aspect-based sentiment classification with multi-attention network. Neurocomputing. 2020;**388**:135-143

[39] Chen Y, Zhuang T, Guo K. Memory network with hierarchical multi-head attention for aspect-based sentiment analysis. Applied Intelligence. 2021; **51**(7):4287-4304

[40] Yuming Lin YF, Li Y, Cai G, Zhou A. Aspect-based sentiment analysis for online reviews with hybrid attention networks. World Wide Web. 2021; **24**(4):1215-1233

[41] Liao W, Zeng B, Yin X, Wei P. An improved aspect-category sentiment analysis model for text sentiment analysis based on roberta. Applied Intelligence. 2021;**51**(6):3522-3533

[42] Kaur R, Kautish S. Multimodal sentiment analysis: A survey and comparison. Research Anthology on Implementing Sentiment Analysis Across Multiple Disciplines. IGI Global. 2022. pp. 1846-1870

[43] Soleymani M, Garcia D, Jou B, Schuller B, Chang S-F, Pantic M. A survey of multimodal sentiment analysis. Image and Vision Computing. 2017;**65**: 3-14

[44] You Q, Luo J, Jin H, Yang J. Cross-modality consistent regression for joint visual-textual sentiment analysis of social multimedia. In: Proceedings of the Ninth ACM International Conference on Web Search and Data Mining. New York: ACM; 2016. pp. 13-22

[45] You Q, Cao L, Jin H, Luo J. Robust visual-textual sentiment analysis: When attention meets tree-structured recursive neural networks. In: Proceedings of the 24th ACM International Conference on Multimedia. New York: ACM; 2016. pp. 1008-1017

[46] Nan X. Analyzing multimodal public sentiment based on hierarchical semantic attentional network. In: 2017 IEEE International Conference on Intelligence and Security Informatics (ISI). Beijing, China: IEEE; 2017. pp. 152-154

[47] Xu N, Mao W. Multisentinet: A deep semantic network for multimodal sentiment analysis. In: Proceedings of the 2017 ACM on Conference on Information and Knowledge Management. New York: ACM; 2017. pp. 2399-2402

[48] Xu N, Mao W, Chen G. A co-memory network for multimodal sentiment analysis. In: The 41st International ACM SIGIR Conference on Research & Development in Information Retrieval. New York: ACM; 2018. pp. 929-932

[49] Yang X, Feng S, Wang D, Zhang Y. Image-text multimodal emotion classification via multi-view attentional network. IEEE Transactions on Multimedia. 2020;**23**:4014-4026

[50] Cheema GS, Hakimov S, Müller-Budack E, Ewerth R. A fair and comprehensive comparison of multimodal tweet sentiment analysis methods. In: Proceedings of the 2021 Workshop on Multi-Modal Pre-Training for Multimedia Understanding. New York: ACM; 2021. pp. 37-45

[51] Masi I, Tran AT, Leksut JT, Hassner T, Medioni G. Do we really need to collect millions of faces for effective face recognition? In: Computer Vision. Cham: Springer; 2016. pp. 579-596

[52] Bahdanau D, Cho K, Bengio Y. Neural machine translation by jointly learning to align and translate. In: International Conference on Learning Representations. New York: ACM; 2014. pp. 46-57

[53] Campagna G, Foryciarz A, Moradshahi M, Lam MS. Zero-Shot Transfer Learning with Synthesized Data for Multi-Domain Dialogue State Tracking. 2020

[54] Chen J, Zhang R, Mao Y, Xu J. Parallel interactive networks for multi-domain dialogue state generation. In: Proceedings of the 2020 Conference on Empirical Methods in Natural Language Processing (EMNLP). Toronto: ACL; 2020. pp. 17-26

[55] Chen H, Liu X, Yin D, Tang J. A survey on dialogue systems: Recent advances and new frontiers. Acm Sigkdd Explorations Newsletter. 2017;**19**(2): 25-35

[56] Gliwa B, Mochol I, Biesek M, Wawer A. Samsum corpus: A human-annotated dialogue dataset for abstractive summarization. In: Proceedings of the 2nd Workshop on New Frontiers in Summarization. New York: ACM; 2019. pp. 38-49

[57] Wen TH, Gasic M, Kim D, Mrksic N, Su PH, Vandyke D, et al. Stochastic language generation in dialogue using recurrent neural networks with convolutional sentence reranking. In: Proceedings of the 16th Annual Meeting of the Special Interest Group on Discourse and Dialogue. Toronto: ACL; 2015. pp. 275-284

[58] Wen TH, Gasic M, Mrksic N, Rojas-Barahona LM, Su PH, Ultes S, et al. Conditional generation and snapshot learning in neural dialogue systems. 2016

[59] Wen TH, Vandyke TH., Mrksic N, Gasic M, Rojas-Barahona LM, Su PH, et al. A network-based end-to-end

trainable task-oriented dialogue system. 2016

[60] Williams J. Multi-domain learning and generalization in dialog state tracking. In: Proceedings of the SIGDIAL 2013 Conference. Toronto: ACL; 2013. pp. 433-441

[61] Williams JD, K. Asadi, G. Zweig. Hybrid code networks:practical and efficient end-to-end dialog control with supervised and reinforcement learning. In: Proceedings of the 55th Annual Meeting of the Association for Computational Linguistics. Toronto: ACL; 2017. pp. 665-677

[62] Tamar A, Yi W, Thomas G, Levine S, Abbeel P. Value iteration networks. In: Twenty-Sixth International Joint Conference on Artificial Intelligence, New York: ACM; 2017. pp. 246-257

[63] Loni B. A survey of state-of-the-art methods on question classification. In: Proceedings of the 7th Workshop on Ph.D Students. New York: ACM; 2011

[64] Tao C, Mou L, Zhao D, Rui Y. Ruber: An unsupervised method for automatic evaluation of open-domain dialog systems. 2017

[65] Tao C, Wu W, Xu C, Hu W, Yan R. One time of interaction may not be enough: Go deep with an interaction-over-interaction network for response selection in dialogues. In: Proceedings of the 57th Annual Meeting of the Association for Computational Linguistics, Toronto: ACL; 2019. pp. 189-197

[66] Tran VK, Nguyen LM. Semantic Refinement Gru-Based Neural Language Generation for Spoken Dialogue Systems. Singapore: Springer; 2017

[67] Tur G, Hakkani-Tur D, Heck L. What is left to be understood in atis? In: Spoken Language Technology Workshop (SLT), New York: IEEE; 2011. pp. 236-247

[68] Lu C, Xiang Z, Cheng C, Yang R, Kai Y. Agent-aware dropout dqn for safe and efficient on-line dialogue policy learning. In: The 2017 Conference on Empirical Methods on Natural Language Processing, Toronto: ACL; 2017. pp. 127-137

Chapter 6

# Deep Learning in Medical Imaging

*Narjes Benameur and Ramzi Mahmoudi*


## Abstract

Medical image processing tools play an important role in clinical routine in helping doctors to establish whether a patient has or does not have a certain disease. To validate the diagnosis results, various clinical parameters must be defined. In this context, several algorithms and mathematical tools have been developed in the last two decades to extract accurate information from medical images or signals. Traditionally, the extraction of features using image processing from medical data are time-consuming which requires human interaction and expert validation. The segmentation of medical images, the classification of medical images, and the significance of deep learning-based algorithms in disease detection are all topics covered in this chapter.


**Keywords:** deep learning, medical imaging, segmentation, classification, diagnosis

## 1. Introduction

Recently, artificial intelligence (AI) is considered as a revolution across the medical field and one of the main factors of this AI revolution is deep learning (DL). The origin of DL and neural networks dates back to 1950. Yet, with the introduction of medically annotated big data, necessary for training, and the availability of high-performance computing, the recent years seem to mark a turning point for DL in medical imaging.

Accordingly, this branch of AI is recently applied to several healthcare problems such as computer-aided diagnosis, disease identification, image segmentation and classification, etc. Unlike classical tools, the powerful key of DL derives from the ability to automatically learn complex features without the need for human interaction. Nevertheless, many challenges still exist in medical health including privacy and heterogeneity of datasets. In this chapter, we will survey the application of DL in clinical imaging, and we will highlight the main challenges and future directions of this tool.

## 2. Deep learning-based segmentation in medical imaging

Deep learning algorithms were used in many medical applications to solve problems with segmentation, image classification, and pathology diagnosis. The manual segmentation process is time-consuming for radiologists because it is typically done slice by slice. Furthermore, segmentation results are susceptible to inter and interobserver variability. To address these limitations, several approaches based on active contour, level set, and statistical shape modeling [1–3] have been proposed to segment the extent of various pathologies or anatomical geometries. All of the methods mentioned above, however, are still semi-automated and require human interaction [4].

With the advent of DL, a fully automated segmentation of serial medical images is become possible in a few seconds. Several studies in the literature reported that segmentation algorithms based on AI outperformed the other classical models [5, 6]. Convolutional neural networks (CNNs) are the most used architecture to segment medical images. It consists of reducing the spatial dimensionality of the original image data through a series of the network layers by performing convolution and pooling operations. Other DL architectures were also proposed for this task such as deep neural network (DNN), artificial neural network (ANN), fully convolutional network (FCN), ResNet-50, and VGGNet-16 [7–10]. **Figure 1** describes the tasks involved in segmenting cardiac images for various imaging modalities.

The success of DL-based medical image segmentation inspired other studies to reevaluate the traditional approaches to image segmentation and incorporate DL models into their work. Many factors have facilitated the increased use of DL. Among them, we can note the availability of medical data and the evolution of graphics processors' performances.

Each year, large, annotated datasets were published online. These data were collected during many challenges such as medical segmentation decathlon and medical image computing and computer aided interventions (MICCAI). **Table 1** summarizes the largest medical images datasets available online.

Segmentation based on DL were applied in different field of medical imaging [12–14]. In cardiac MRI, several DL models were used to delineate the contours of the myocardium which represent a crucial step to compute useful clinical parameters for the evaluation of cardiac function [15]. DL was also applied for the segmentation of different types and stage of cancer. For breast cancer, the data include mammography, ultrasound, and MRI images [16–18]. Other DL architectures were also proposed in the literature to segment cervical cancer based on Magnetic Resonance Imaging (MRI), computed tomography (CT), and positron emission tomography (PET) scan

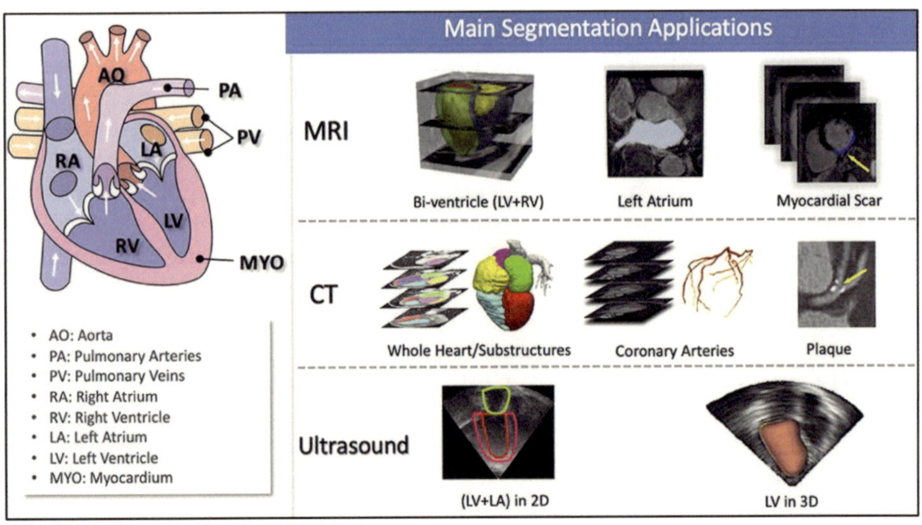


**Figure 1.**
*Overview of cardiac image segmentation tasks for different imaging modalities [11].*

| Dataset | Target | Modality | Source |
|---|---|---|---|
| Kaggle | Various diseases | X-rays, MRI and CT | Google LLC *https://www.kaggle.com/* |
| NIH Image Gallery | Various diseases | X-rays, MRI, CT, PET. | National Institutes of Health (NIH) *https://www.flickr.com/photos/nihgov/* |
| ImageNet | Cancer, diabetes, and Alzheimer's disease. | Genomic data | AI researchers https://www.image-net.org/ |
| Google Dataset Search | Various diseases | X-rays, MRI, CT, PET, echography | Google *https://datasetsearch.research.google.com/* |
| UCI Machine Learning Repository | Various diseases | X-rays, MRI, CT, PET, echography | The National Science Foundation *https://archive.ics.uci.edu/ml/index.php* |
| Stanford Medical ImageNet | Various diseases | X-rays, MRI scans, and CT scans. | Stanford University *https://aimi.stanford.edu/medical-imagenet* |
| Open Images Dataset | Various diseases | All medical Imaging techniques | Google in collaboration with CMU and Cornell Universities *https://storage.googleapis.com/openimages/web/index.html* |
| Cancer Imaging Archive | Cancer | Images from a variety of cancer types | National Cancer Institute (NCI) *https://www.cancerimagingarchive.net/access-data/* |
| Alzheimer's Disease Neuroimaging Initiative | Alzheimer's disease | brain scans and related data from MRI | Foundation for the National Institutes of Health *https://adni.loni.usc.edu/* |
| The Microsoft COCO dataset | Various diseases | All medical Imaging techniques | Microsoft *https://cocodataset.org/#home* |
| MIAS dataset | Various diseases | Mammographic images | Organization of UK research groups *https://www.kaggle.com/datasets/kmader/mias-mammography* |

**Table 1.**
*Medical images datasets available online.*

data [19]. Zhao et al. [20] proposed a new model of DL that combined U-net with progressive growing of U-net+ (PGU-net) for automated segmentation of cervical nuclei. In their study, they reported a segmentation accuracy of 92.5%. Similarly, Liu et al. [21] applied a modified U-net model on CT images for clinical target volume delineation in cervical cancer. In their proposed architecture, the encoder and decoder components were replaced with dual path network (DPN) components. The mean dice similarity coefficient (DSC) and the Hausdorff distance (HD) values of the model were 0.88 and 3.46 mm.

Although image segmentation based on DL facilitates the detection, characterization, and analysis of different lesions in medical images, it still suffers from several limitations. First, the problem of missing border regions in medical images should be considered [22]. Furthermore, the imbalanced data available online could significantly affect the segmentation performances. In medical imaging, the collection of balanced data is challenging since images related to controls are largely available compared to those associated with different pathologies. As a result, some models have been proposed to mitigate this problem. These models include convolutional autoencoders [23] and generative adversarial networks (GAN) [24]. The concept is based on the extraction of information from original images and to generate a similar dataset image based on linear transformation, e.g., reflection, rotation, translation.

## 3. Deep learning-based classification in medical imaging

Additionally, DL have demonstrated its superiority in the categorization of medical images, more notably in the distinction of various disorders. The extraction of key features is a step in the classification process that produces a model that can categorize a picture into multiple classes. To extract features using color or texture, several classical classifications have been dressed in the literature [25–27]. Support vector machines (SVMs), logistic regression, closest neighbors, etc. can be mentioned among them. These systems must, however, cope with other challenging problems related to medical imaging. First, the presence of artifacts in medical images may make it more difficult to categorize. Because of this, pre-processing is crucial to improving image quality. The second problem is the complexity of medical content captured by many modalities. The classification of medical images is extremely difficult because each modality has distinct characteristics.

Recently, several researchers used DL for the medical classification task and the results proved the accuracy of their models in comparison with the traditional machine learning approaches [28]. Deep learning's key benefit is its ability to quickly distinguish between various structures in images without the need for manual feature extraction. Recent DL architectures also have the capacity to incorporate a variety of features gathered from many modalities to produce an effective classifier.

Yadav and Jadhav [29] used a DL algorithm based on the transfer learning of VGG16 to classify pneumonia from chest X rays' images. In their study, they showed that the VGG16 outperformed the classical method based on SVM. The accuracy was 0.923 for VGG16 vs. 0.776 for SVM. Similarly, Xu et al. [30] tested a deep CNN in histopathology images to extract new features for the classification of colon cancer. Lai et al. [31] proposed a new architecture that combines coding network with multilayer perceptron (CNMP) with other features extracted from deep CNN.

In their study, they showed an accuracy of 90.2%. Although DL achieved high performance in the classification of medical images, it still suffers from numerous limitations. The major challenge is the reduced number of annotated data needed for the classification of medical images. Labeling data require the intervention of experienced radiologists. A few solutions have been proposed to resolve this issue. Pujitha and Sivaswamy [32] proposed a crowd-sourcing and synthetic image generation for training deep neural net-based lesion detection. In their study, they used color fundus retinal and they proved that crowd-sourcing improves the area under the curve (AUC) by 25%. The generative adversarial networks (GAN) is also another source of generating synthetic images with annotations. Aljohan and Alharbe [33] proposed a new GAN to generate synthetic medical images with the corresponding annotations from different medical modalities. The classification of medical images based on DL has shown good results. However, there are still several issues in medical image processing that need to be addressed with the different DL architectures.

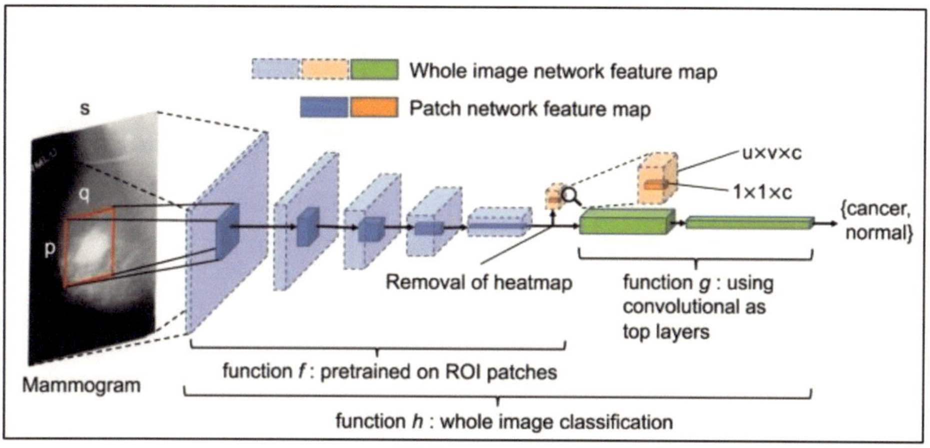


**Figure 2.**
*Deep learning for the screening of breast cancer [37].*

## 4. Disease diagnosis based on deep learning

Early and precise diagnosis is crucial for the treatment of different diseases and for the estimation of a severity grade. The use of DL for the diagnosis of diseases is a dynamic research area that attracts several researchers worldwide. In fact, DL architectures have been applied to some specific pathologies such as cancer, heart disease, diabetes, and Alzheimer's disease [34, 35]. The increasing number of medical imaging dataset led different researchers to use deep learning models for the diagnosis of different diseases.

DL algorithms have proven their performances in the prediction and diagnosis of cancer diseases. The availability of images derived from MRI, CT, mammography, and biopsy helped several researchers to use these data for early cancer detection. The analysis of cancer images includes the detection of tumor area, the classification of different cancer stages, and the extraction of different characteristics for tumors [36].

Recently, Shen et al. [37] used a modified version of CNNs for the screening of breast cancer using mammography data. The outcomes of their study showed an AUC of 0.95 and a specificity of 96.1%. A CNN was also applied for the classifications of different kinds of cancer and the detection of carcinoma. **Figure 2** depicts the entire image categorization process for breast cancer screening using DL architecture.

Alanazi et al. [38] applied the transfer DL model to detect brain tumor in the early stage by using various types of tumor data. Furthermore, another study used a 3D deep CNN to assess the glioma grade (low or high-grade glioma). In their study, they reported an accuracy of 96.49% [39]. Compared to the classical algorithms, the different studies proved the efficiency of DL in the prediction and analysis of cancer. However, bigger medical data available online are needed for more adequate validation.

## 5. Conclusion

As has been shown, using medical image processing techniques in clinical practice is crucial for determining if a patient has a particular disease or not. The field of medical imaging has been transformed by AI and DL, which enable more precise and automatic feature extraction from medical data. DL has been used to address a variety of healthcare issues, including image segmentation and classification, disease detection, computer-aided diagnosis, and the learning of complex features without human interaction. Despite the advances made, many challenges still exist in medical health including privacy and heterogeneity of datasets.

## Conflict of interest

The authors declare no conflict of interest.


## Author details

Narjes Benameur[1] and Ramzi Mahmoudi[2,3]*

1 Laboratory of Biophysics and Medical Technology, Higher Institute of Medical Technologies of Tunis, University of Tunis el Manar, Tunis, Tunisia

2 Faculty of Medicine, Laboratory of Technology and Medical Imaging, University of Monastir, Monastir, Tunisia

3 Gaspard-Monge Computer Science Laboratory, A3SI, ESIEE Paris, Gustave Eiffel University, France

*Address all correspondence to: ramzi.mahmoudi@esiee.fr

## References

[1] Pohl KM, Fisher J, Kikinis R, Grimson WEL, Wells WM. Shape based segmentation of anatomical structures in magnetic resonance images. Computer Visual Biomedical Image Application. 2005;**3765**:489-498. DOI: 10.1007/11569541_49

[2] Chen X, Williams BM, Vallabhaneni SR, Czanner G, Williams R, Zheng Y. Learning active contour models for medical image segmentation. In: IEEE/CVF Conference on Computer Vision and Pattern Recognition (CVPR). Long Beach, CA, USA; 2019. pp. 11624-11632. DOI: 10.1109/CVPR.2019.01190

[3] Swierczynski P, Papież BW, Schnabel JA, Macdonald C. A level-set approach to joint image segmentation and registration with application to CT lung imaging. Computerized Medical Imaging and Graphics. 2018;**65**:58-68

[4] Gao Y, Tannenbaum A. Combining atlas and active contour for automatic 3d medical image segmentation. Proceedings of the IEEE International Symposium Biomedical Imaging. 2011;**2011**:1401-1404

[5] Kim M, Yun J, Cho Y, Shin K, Jang R, Bae HJ, et al. Deep learning in medical imaging. Neurospine. 2019;**16**(4):657-668

[6] Vaidyanathan A, van der Lubbe MFJA, Leijenaar RTH, van Hoof M, Zerka F, Miraglio B, et al. Deep learning for the fully automated segmentation of the inner ear on MRI. Scientific Reports. 2021;**11**(1):2885

[7] Zadeh Shirazi A, McDonnell MD, Fornaciari E, Bagherian NS, Scheer KG, Samuel MS, Yaghoobi M, Ormsby RJ, Poonnoose S, Tumes DJ, Gomez GA. A deep convolutional neural network for segmentation of whole-slide pathology images identifies novel tumour cell-perivascular niche interactions that are associated with poor survival in glioblastoma.

[8] Cai L, Gao J, Zhao D. A review of the application of deep learning in medical image classification and segmentation. Annals of Translational Medicine. 2020;**8**(11):713. DOI: 10.21037/atm.2020.02.44

[9] Malhotra P, Gupta S, Koundal D, Zaguia A, Enbeyle W. Deep neural networks for medical image segmentation. Journal of Healthcare Engineering. 2022;**200**:9580991. DOI: 10.1155/2022/9580991

[10] Alsubai S, Khan HU, Alqahtani A, Sha M, Abbas S, Mohammad UG. Ensemble deep learning for brain tumor detection. Frontiers in Computer Neuroscience. 2022;**16**:1005617. DOI: 10.3389/fncom.2022.1005617

[11] Chen C, Qin C, Qiu H, Tarroni G, Duan J, Bai W, et al. Deep learning for cardiac image segmentation: A review. Frontiers in Cardiovascular Medicine. 2020;**7**:25. DOI: 10.3389/fcvm.2020.00025

[12] Hesamian MH, Jia W, He X, Kennedy P. Deep learning techniques for medical image segmentation: Achievements and challenges. Journal of Digital Imaging. 2019;**32**(4):582-596. DOI: 10.1007/s10278-019-00227-x

[13] Fu Y, Lei Y, Wang T, Curran WJ, Liu T, Yang X. A review of deep learning based methods for medical image multi-organ segmentation. Physica Medica. 2021;**85**:107-122

[14] Bangalore Yogananda CG, Shah BR, Vejdani-Jahromi M, Nalawade SS, Murugesan GK, Yu FF, et al. A fully automated deep learning network for brain tumor segmentation. Tomography. 2020;**6**(2):186-193

[15] Wang Y, Zhang Y, Wen Z, Tian B, Kao E, Liu X, et al. Deep learning based fully automatic segmentation of the left ventricular endocardium and epicardium from cardiac cine MRI. Quantitative Imaging in Medicine and Surgery. 2021;**11**(4):1600-1612

[16] Abdelrahman A, Viriri S. Kidney tumor semantic segmentation using deep learning: A survey of state-of-the-art. Journal of Imaging. 2022;**8**(3):55

[17] Yue W, Zhang H, Zhou J, Li G, Tang Z, Sun Z, et al. Deep learning-based automatic segmentation for size and volumetric measurement of breast cancer on magnetic resonance imaging. Frontiers in Oncology. 2022;**12**:984626

[18] Caballo M, Pangallo DR, Mann RM, Sechopoulos I. Deep learning-based segmentation of breast masses in dedicated breast CT imaging: Radiomic feature stability between radiologists and artificial intelligence. Computers in Biology and Medicine. 2020;**118**:103629

[19] Yang C, Qin LH, Xie YE, Liao JY. Deep learning in CT image segmentation of cervical cancer: A systematic review and meta-analysis. Radiation Oncology. 2022;**17**(1):175

[20] Zhao Y, Rhee DJ, Cardenas C, Court LE, Yang J. Training deep-learning segmentation models from severely limited data. Medical Physics. 2021;**48**(4):1697-1706

[21] Liu Z, Liu X, Guan H, Zhen H, Sun Y, Chen Q, et al. Development and validation of a deep learning algorithm for auto-delineation of clinical target volume and organs at risk in cervical cancer radiotherapy. Radiotherapy and Oncology. 2020;**153**:172-179

[22] Zambrano-Vizuete M, Botto-Tobar M, Huerta-Suárez C, Paredes-Parada W, Patiño Pérez D, Ahanger TA, et al. Segmentation of medical image using novel dilated ghost deep learning model. Computational Intelligence and Neuroscience. 2022;**2022**:6872045

[23] Gondara L. Medical image denoising using convolutional denoising autoencoders. In: 2016 IEEE 16th International Conference on Data Mining Workshops (ICDMW). Barcelona, Spain; 2016. pp. 241-246. DOI: 10.1109/ICDMW.2016.0041

[24] Gulakala R, Markert B, Stoffel M. Generative adversarial network based data augmentation for CNN based detection of Covid-19. Scientific Reports. 2022;**12**:19186

[25] Shukla P, Verma A, Verma S, Kumar M. Interpreting SVM for medical images using Quadtree. Multimedia Tools and Applications. 2020;**79**(39-40):29353-29373

[26] Tchito Tchapga C, Mih TA, Tchagna Kouanou A, Fozin Fonzin T, Kuetche Fogang P, Mezatio BA, et al. Biomedical image classification in a big data architecture using machine learning algorithms. Journal of Healthcare Engineering. 2021;**2021**:9998819

[27] Rashed BM, Popescu N. Critical analysis of the current medical image-based processing techniques for automatic disease evaluation: Systematic literature review. Sensors (Basel). 2022;**22**(18):7065

[28] Puttagunta M, Ravi S. Medical image analysis based on deep learning

approach. Multimedia Tools and Applications. 2021;**80**(16):24365-24398

[29] Yadav SS, Jadhav SM. Deep convolutional neural network based medical image classification for disease diagnosis. Journal of Big Data. 2019;**6**:113

[30] Xu Y, Jia Z, Wang LB, Ai Y, Zhang F, Lai M, et al. Large scale tissue histopathology image classification, segmentation, and visualization via deep convolutional activation features. BMC Bioinformatics. 2011;**18**(1):281

[31] Lai Z, Deng H. Medical image classification based on deep features extracted by deep model and statistic feature fusion with multilayer perceptron. Computational Intelligence and Neuroscience. 2018;**2018**:2061516

[32] Pujitha AK, Sivaswamy J. Solution to overcome the sparsity issue of annotated data in medical domain. CAAI Transactions on Intellectual Technology. 2018;**3**:153-160

[33] Aljohani A, Alharbe N. Generating synthetic images for healthcare with novel deep Pix2Pix GAN. Electronics. 2022;**11**(21):3470. DOI: 10.3390/electronics11213470

[34] Kumar Y, Koul A, Singla R, Ijaz MF. Artificial intelligence in disease diagnosis: A systematic literature review, synthesizing framework and future research agenda. Journal of Ambient Intelligence and Humanized Computing. 2022;**2022**:1-28

[35] Ibrahim A, Mohamed HK, Maher A, Zhang B. A survey on human cancer categorization based on deep learning. Frontiers in Artificial Intelligence. 2022;**5**:884749. DOI: 10.3389/frai.2022.884749

[36] Tran KA, Kondrashova O, Bradley A, et al. Deep learning in cancer diagnosis, prognosis and treatment selection. Genome Medicine. 2021;**13**:152. DOI: 10.1186/s13073-021-00968-x

[37] Shen L, Margolies LR, Rothstein JH, et al. Deep learning to improve breast cancer detection on screening mammography. Scientific Reports. 2019;**9**:12495. DOI: 10.1038/s41598-019-48995-4

[38] Alanazi MF, Ali MU, Hussain SJ, Zafar A, Mohatram M, Irfan M, et al. Brain tumor/mass classification framework using magnetic-resonance-imaging-based isolated and developed transfer deep-learning model. Sensors (Basel). 2022;**22**(1):372. DOI: 10.3390/s22010372

[39] Mzoughi H, Njeh I, Wali A, Slima MB, BenHamida A, Mhiri C, et al. Deep multi-scale 3D Convolutional Neural Network (CNN) for MRI gliomas brain tumor classification. Digital Imaging. 2020;**33**(4):903-915. DOI: 10.1007/s10278-020-00347-9